A
Well-Tempered
Mind

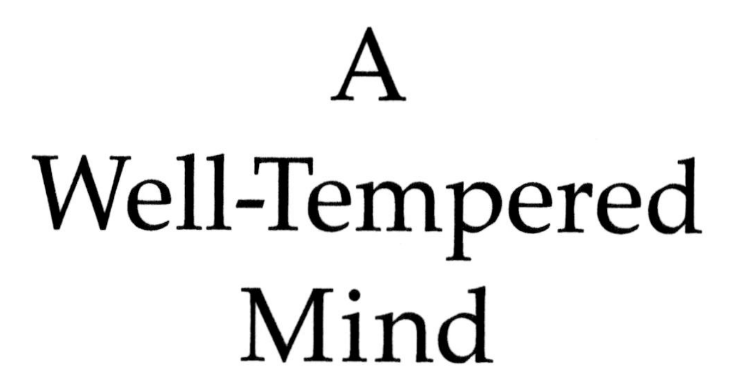

A Well-Tempered Mind

Using Music to Help Children Listen and Learn

PETER PERRET AND JANET FOX
FOREWORD BY MAYA ANGELOU

NEW YORK • WASHINGTON, D.C.

Published by Dana Press
New York | Washington, D.C.

The Dana Foundation
745 Fifth Avenue, Suite 900
New York, NY 10151

900 15th Street, NW
Washington, DC 20005

ISBN: (cloth) 1-932594-03-5
ISBN: (paper) 1-932594-08-6

Library of Congress Cataloging-in-Publication Data

Perret, Peter, 1941-

A well-tempered mind : using music to help children listen and learn / Peter Perret & Janet Fox ; drawings by Susette Louise Sides.

p. cm.

Includes bibliographical references and index.

ISBN 1-932594-03-5 (hardcover : alk. paper)

1. Music—Psychological aspects. 2. Music in education—North Carolina—Winston-Salem. 3. Music, Influence of. 4. Learning, Psychology of. I. Fox, Janet, 1944- II. Title.

ML3830.P375 2004

372.87—dc22

2004006041

Drawings by Susette Louise Sides
Design by Jeff Hall, Ion Graphic Design Works
Cover design by Ellen Davey, illustration by Susette Sides

TO THE BOLTON/ABES QUINTETS 1994–2004
Robert, Lisa, Bob, Eileen, Kristine, Steve, Sonja, Jon, Cara,
Debi, James, Kendall, Kathy, Amber, Carol, and Tim

Boldly going where no woodwind quintet
had gone before!

Special thanks to Susette Sides
for her wonderful drawings.

CONTENTS

FOREWORD

Recognized or not, directly or obliquely, music is a major factor in the life of every human being. The ear is attuned to hearing melody from birth, whether the melody is that of the mother's lullaby or the tune of languages spoken. Spirits are increased by the gift of music.

When a music maker belongs in a community and identifies with that community, the community is increased by the gift, the music maker is increased by the gift, and even the music itself is increased.

Maestro Peter Perret and Janet Fox here illumine the power of music to uplift and even transform lives. Perret and Fox are aware of the importance of music in the community and in the world.

In this book, they prove the efficacy of music in education. Perret's success with third graders is so appealing that the most cynical disbeliever must be won over and then agree with John Adams's statement in a letter to Abigail Adams, 1780, "I must study politics and war that my sons may have liberty to study mathematics and philosophy, geography, natural history, naval architecture, navigation, commerce and agriculture, in order to give their children a right to study [many arts including] music."

I pray the gift of this book, along with the gift of music, will herald the return of art in the classroom. The children need that and so does our world.

Maya Angelou

INTRODUCTION

> It is not easy to determine the nature of music, or why any one should have a knowledge of it...The first question is whether music is or is not to be a part of education.
>
> —Aristotle, *Politics*, translated by Benjamin Jowett

Everybody thinks music is important, but nobody knows exactly why. As with so many issues in modern cognitive neuroscience, Aristotle's original formulation of this problem is instructive, and we would do well to follow his lead. He helps us understand the stage on which the drama of research on music and mind will be played—but in the 23 centuries since Aristotle, it can fairly be said that only in the last two decades have researchers even moved onto that stage. The excitement is all still ahead. But before considering what lies ahead, let us take advantage of Aristotle's insight.

It is remarkable in the first place that Aristotle's treatment of music is found in his *Politics*, rather than his *Treatise on the Soul*. (The latter is not a religious discussion but an insightful account of the biology and psychology of basic bodily processes such as nutrition, sensation, action, and thought.) For Aristotle, music is less about the philosophy or science of mind and more about a particularly central function of government—the education of youth, for the sake of responsible citizenship. It is in that context that he considers alternative explanations of the value or power of music: relaxation (a good thing in itself); character building (analogous to the way exercise develops the body); and cultivation of the mind (which is accomplished in a structured kind of leisure, free from the cares of work). He concludes that music to some extent aids all three, but it especially cultivates the mind, during leisure. Leisure is for Aristotle an intellectual experience and is distinguished from relaxation, which is pure respite from the cares of work. All this rests on a more basic foundation, a simple but powerful insight: music, he says, is not work.

For Aristotle, we miss the power of music if we make it into work, either through paid performance, where the worker is the musician, or by making music an object of study—that is, a "subject" to be taught and learned. To be sure, music is not to be only passive listening; instead, explicit training in the production of music, both vocal and instrumental, is necessary. This training, however, is to equip the student to listen to music constructively and thereby to derive the maximum intellectual benefit. But the insight is more positive than negative: a "work" ethic of music is to be avoided if we are to gain the affirmative value of music as leisure. In specific contrast to reading and writing, which Aristotle considers valuable because of the advantage to society of the explicit knowledge they communicate, music accomplishes nothing external and carries no content. Instead, it is to be enjoyed for its own sake, which is an intellectual benefit only to the individual. In Aristotle's terminology, music is neither useful nor necessary, but it is noble.

Of course, the question before us is how that intellectual benefit to the individual is conferred. Here Aristotle is clear: music, unlike the simple perception of auditory or visual objects, imitates mental and emotional states so powerfully that they can actually be induced in the listener. To use contemporary scientific terms, the states are "mapped"—from mind to music and back again to mind. More important, the states thus induced, while they can be labeled with words, are without verbal content. Still more important, these states are "intellectual" in every sense. The agenda for contemporary cognitive neuroscience research on music, then, is this: How is this kind of intellect transmissible apart from verbal content?

If asked, Aristotle would certainly have appealed to mental "forms" as the vehicle for such transmission. His entire philosophy and psychology is based on the notion that intellect consists of recognizing the form—one might say, the "mental shape"—of things. For example, Aristotle points out that we have no difficulty mapping emotions such as anger and grief into music, so that those who hear can readily discern and experience the emotional state being conveyed. A form, without content, is communicated. A mental map is drawn by the music, and if that map is mentally traveled, an emotion or mental state is experienced. Nor is emotion the only communicable state; various aspects of intellect, especially those that are difficult to reduce to language, also are equally well-mapped and transmissible. Thus, if cognitive neuroscience lets Aristotle guide its investigations of music, it will look at the mapmaking functions of the brain, particularly those involved in making cognitively interesting maps that represent sequences of sounds.

We miss the real genius of Aristotle's treatment, however, if we overlook his central insight: questions about music are best conceived as questions about education. The essence of music, as form without content, is philosophically interesting, and it provides a promising avenue of research in cognitive neuroscience. How forms are registered in the brain, and how emotional or mental states can be induced by those registered forms, is a fair question that the sciences can properly investigate, to good effect. It is, however, in education that we can most profitably study these questions. Here we no longer speak about the brain but instead about how a particular child's music experience influences his or her intellectual growth and achievement. If Aristotle was right that music is mainly an educational question, then it may be that the fastest progress in the cognitive neuroscience of music will depend on studies of children learning. How music "works" in the classroom, and what outcomes it produces, may have the most to teach us about how music "works" in the brain.

Since the late 1980s the efforts to describe the effect of music on the brain have been divided into "inborn" and "acquired" explanations. Those emphasizing music as an inborn skill have argued that the very structure and function of the human brain predisposes it to music, not unlike the consensual predisposition of the human brain toward speech and language. The contrary view has stressed learned factors, suggesting that the human capacity to analyze and "understand" music is no different from the capacity to analyze any other class of events. In this view, music is not a "special" endowment of the brain. This debate is reminiscent of a highly similar discussion, during the past half century, regarding inborn versus acquired features of speech and language. However, while a consensus has emerged that some aspects of the ability to produce speech and language are properly considered innate or "hard-wired," no such consensus has yet arisen regarding music: the available neuroscientific evidence is still too preliminary. Consequently, two theoretically distinct options for how music affects the brain remain possible: (1) that music experience elicits an inborn skill that then becomes a tool for broader nonmusical reasoning—similar to the way language experience elicits a skill that equips children with verbal tools they can use for abstract reasoning; or (2) that music experience simply contains abstract properties that reason can extract and then use to advantage in solving nonmusical problems that contain those same properties.

Following Aristotle, however, we linger only momentarily on the brain before moving to the classroom, where the greater truth is to be found. Whatever the final verdict of neuroscience on music as a special property of the human brain, a series of research findings has begun to suggest an academically beneficial impact of music instruction. While this series is not wide-ranging or extensive, there are replicated findings suggesting that if students themselves actively engage in certain kinds of music training (thereby learning to produce, rather simply perceive, the music), then at least in the short term they may show improvement in certain kinds of problem-solving skills. One of the recent of these, with which I am personally familiar, is the doctoral dissertation of Shirley Bowles, at the Indiana University in Pennsylvania, in 2003, described later in this book. As her study showed, in a prospective randomized design, Peter Perret's Bolton curriculum of interactive music instruction, delivered in 18 weekly half-hour sessions in a single semester in kindergarten, first, second, or third grade, effected some significant improvements in children's reasoning skills, particularly in phonemic awareness—the latter widely accepted as the necessary precursor for learning to read.

Phonemic awareness is a purely auditory-oral skill in isolating individual sounds. For example, if you ask someone to say "cow" without the "k" sound, the answer is "ow." ("K" is the consonant sound made only with the throat, not the syllable "kuh" or "kay.") If one asks this of a young child, say a five-year-old, prior to any learning of letters, one sometimes sees the child's eyes move left and right in response to the question. This suggests what is later confirmed by the neuroscience investigations: even this purely auditory skill may require some degree of cross-modal association, or "visualizing," of the word sounds. As that skill develops, it allows children to begin matching (we prefer *mapping*) letters to the individual sounds they make. This cross-modal mapping reminds us of Aristotle's notion that music is a consummate mapping skill—a process whereby an entire mental state can be translated into sound, communicated, and retranslated to experience by the listener. Does music stimulate the brain mechanisms of cross-modal association or mapping, thereby equipping the brain for phonemic awareness? In my opinion, yes, but as we should have expected from Aristotle, the classroom science is further advanced than the laboratory science, though they already overlap.

This constructive book puts the matter exactly where Aristotle wanted it, in a proven technique for music instruction for children. As in many circumstances across intellectual history, something that works can be put to work before science fully understands why it works. The Bolton curriculum, I can now say from firsthand experience as a research colleague of Peter Perret and a mentor of Shirley Bowles, has proved effective for enhancing cognitive skills, including the skills that support learning to read. When children listen for faint sounds in the near silence, when they imagine pictures of sounds going up and down on a page, when they discuss why longer woodwind instruments might make deeper sounds than shorter ones, and when they learn to listen for the complex forms in "real" music, not only are they learning what cognitive neuroscience considers essential for intellect, but they also are teaching cognitive neuroscience how the brain works. After 23 centuries of waiting, the prospect of a neuroscience of music built on classroom experience is at hand.

As prospects go, this one is especially vivid—which is to say that it raises especially clear questions. What, exactly, is being taught and learned? The best short answer seems to be "how to make maps in the mind." Like all things learned, this skill has a discipline. Making maps requires an organized set of representational codes. For example, speed is represented by how long any note endures before the next note; direction or mood, by pitch (high versus low tones); comfort, by persisting recurrence of simple melodic figures resolving to major chords; and interest, by variations in the degree to which one note or series of notes predicts the next. Following Aristotle, we furthermore expect that while much of this is learned and appreciated simply by listening, the full scope of the learning is not realized until children learn to make their own music—their own compositions as well as those of others. Doing so trains the mapmaking facilities (as does athletics), making them not only more flexible but also more precise. That immediately raises the next question.

How does music learning affect other intellectual competencies? Again, the short answer might be "by encouraging the appreciation of forms apart from content." All discourse, whatever its content, has fast and slow parts, highs and lows, and recurrences and surprises—to mention only a few of the forms that music teaches. Children who master such content-free forms are then able fluently to create receptacles into which they receive, interpret, and categorize their experience with any academic or vocational content.

The analogy between vessel-making and music also suggests the answer to another major question: Should music be taught as a skill in its own right, or should it simply be used as a method of teaching specific content—as in learning to sing the alphabet? The former. If music is taught "only" as a vessel for particular content, then by definition the forms remain specific to that content. If, instead, music is taught for itself, generic forms applicable to any content are learned; moreover, the skill of instantly recognizing and manufacturing the forms, when confronted with any new content, is also learned. We must not let this notion be reduced to the oversimplified idea that a child simply learns to put his or her lessons into some kind of musical and perhaps singable mnemonic form. Instead, the forms of music, once learned by experience with music, are also recognizable as the forms of text or of ideas—again, the fast or slow, high or low, and recurrent or surprising features, among many others.

All short answers, including the above, are really hypotheses for further scientific test. These lend themselves readily to that, by a variety of experiments that are the stock-in-trade of cognitive neuroscience. For example, we will need to explore systematically whether the specific type of music (solo melodies versus counterpoint versus harmony, for example) is important, whether all the relevant forms are inherent in any music or whether specific practice with some forms differentially facilitates the recognition of those particular forms in other domains of nonmusical discourse, and whether any or all of the forms require for their mastery that learners become skilled in their production as well as their perception.

It is inherent in science that experiments addressing particular questions, like those above, are never final results, but instead always sequences of results, so that each raises further questions for subsequent experiments. While some questions do get answers, some never do, and some important questions are not even knowable in the early stages of an investigation. For example, it is necessary to amplify Shirley Bowles's experiment (which showed that a music curriculum improved phonemic awareness in children) with many follow-up studies to isolate the particular mechanisms of the effect. Experiments will address the particular features of the curriculum (such as listening for specific sounds, drawing sequences of sounds on paper, or discussing how music is made by instruments) that are responsible for the effect. A laboratory setting, in which these various activities are conducted with randomly chosen groups of children, would probably be required. Whatever the setting, these types of instruction would be preceded and followed by extensive tests of phonemic awareness in its various forms (by asking children to produce a word that is "left over" when a sound is deleted from that word, by asking them to listen to extended speech and indicate when they hear a specific sound, and so on). It could well be that specific features of instruction would influence specific features of phonemic awareness. Only in this way, by patiently conducting the broad range of relevant experiments, will these new findings eventually become reliably well understood and efficiently usable. The very fact that these refined questions are now askable, and answerable, is a tribute to the fact that a new chapter in the science of learning and of thought is before us. These musicians in the classroom have taken us to a whole new continent in the world of brain and mind, a place where the landscapes impress us equally with their vast scope and their astonishing local detail.

My collaborations with Peter Perret have taught me that his hands not only wield the baton and trace the melody but also search the literature and gesture the unfolding of scientific truth. True to the cross-modal nature of his subject, he has become a diligent student of the neurosciences, and this overlap tells me that he is "on to" things that are as fundamental to neuroscience as to music.

Frank B. Wood, Ph.D.
Professor of Neurology
Head, Section of Neuropsychology
Wake Forest University School of Medicine

PREFACE

This is the story of a project that members of the Winston-Salem Symphony and I, their director, designed for an at-risk elementary school in the mid-1990s. It was born of an idea that our formal musical knowledge could be used in specific creative ways to help those children learn. It succeeded beyond our wildest expectations. Measured by the standard state-prescribed tests of reading, writing, and mathematical achievement, the children we worked with did better than expected and better than children who hadn't had their instruction blended with music. That was the case for the first class we worked with and every subsequent class in that school and in another in our school district in which we expanded the program. Schools elsewhere that have used our program as a model have reported similar results. Yet we have not taught the kids to play a single note.

The program began almost exactly a decade ago, in the 1994–95 school year, at Bolton Elementary School in Winston-Salem, North Carolina, and has attracted attention from people from around the country who are interested in music, education, and science. When I speak to groups of educators, parents, musicians, symphony organizations, and scientists about the music program we started, they react with curiosity and excitement.

And curiosity and excitement are what I feel as I read what neuroscientists discover in their desires to unravel the mysteries of the brain. The research in and of itself presents a complicated, often controversial, and certainly ever-changing picture. Scientists are discovering that many different parts of the brain involving many networks are used in processing almost any action, perception, or idea. So it is difficult to show that the effect of music is specific to some, but not other, regions of the brain. Although we saw gratifying improvements in the children's scholastic achievement, I would not be so brave or so brash as to believe that we have yet isolated what particular features of music or its teaching might be responsible for which changes in brain activity. Rather, music may improve cognitive agility—that is, the ability to manage competing types of information and choose which type to apply in a given situation.

What might be more likely is that musical performance, with its demands of score-reading, memory, motor control, coordination, and emotional expressivity, may eventually prove to train a variety of perceptual and executive functions, leading to greater cognitive agility and competence. But it is a narrower question whether learning about music and learning to listen with greater subtlety by itself leads children to greater mental agility.

Multisensory learning—that is, learning by using all the senses, which is certainly part of what we taught—uses competing sources of information and may cause the brain to choose which source to apply and how. Linking these competing (or complementary) sensory inputs may cause the formation of vaster and richer networks. Science is laying the groundwork to explain the quintet's experience. Researchers interested in the arts and how children learn are finding promising signs that improvements such as those I describe in these pages are physiological and not purely psychological. This book is just a beginning.

Distinguished leaders in the field of learning and the brain have suggested to us that other subjects, made equally interesting to children, might have similar potential. Although music may not be the only possible subject, a well-planned program that will interest and engage young children is essential. And while researchers continue their work, the children have a need now, and professionals in many fields can learn something from the story of the quintet.

Ever since I began making presentations on this subject, people have been asking for "the book." So here is the book, written over the past three years with my longtime collaborator, Janet Fox. I am a musician, a symphony conductor, a parent of school-age children, and a person with a lifelong interest in education and in science. Since I began thinking about trying to use music to improve academic achievement, I have been a student of brain science and have avidly read about the research that might help explain the results at Bolton. I have combined my knowledge of music and my investigation of the science to explore how music might build brainpower.

Janet is a professional writer who has specialized in education, the arts, and health and medicine. She is also a mother and grandmother with a keen interest in the development of young minds. And she has been involved in the Bolton project from its inception, having been the Winston-Salem Symphony's public relations director at that time. While the "I" of the narration stands for Peter, both our voices are present.

WHY "WELL-TEMPERED"?

Unlike electronic instruments or compact discs, which will always sound the same, musical instruments can vary their characteristics enormously. For example, players can change how their instruments are tuned. This enables musicians to adjust to local conditions and to play in tune with other musicians. Some orchestra musicians like to joke that they don't need to tune because their instrument was tuned at the factory. In reality, when they practice and play, they tune and retune several times an hour.

The tuning of orchestral instruments can be adjusted in a few seconds, which is what the musicians are doing when the oboe plays an A just before the conductor walks onto the stage at a concert. However, it takes half an hour to tune a harp, and a couple of hours to tune a piano. Tuning a pipe organ is so cumbersome that it is done only every few years, and then at great length and cost.

In the history of music, a profound change in the tuning of instruments occurred at the beginning of the Baroque period. Until then, instruments were tuned to perfect fifths, an interval between notes that eventually leads to a feeling of out-of-tune-ness in keys having many sharps (#'s) or many flats (♭'s). A minuscule splitting of the difference, whereby the fifths were tuned slightly flat, solved the problem, and soon composers began writing music for instruments that were equally tempered—that is, having equal spaces between the notes—in contrast to the mathematically based form of tuning that had been traditional for 2,000 years. Predictably, musicians resisted this change, and many claimed it would never work. Johann Sebastian Bach wrote a series of preludes and fugues, two in every major and minor key, 48 in all, to prove that great music could be produced using the tempered scale. We know that series as *The Well-Tempered Clavier*.

What equal tempering did was to make it possible to play in all keys, relatively in tune, rather than in only a few keys. Equal tempering opened the way for many, many kinds of music that we enjoy today. What was actually a slight change made an enormous difference.

The richness and resonance of the interaction between our musicians and the children in the classroom cannot be fully captured in words or fully accounted for by scientific measurement. It takes no great stretch of the imagination to believe that these interactions are tempering young minds, in positive and far-reaching ways.

It is our hope that all those—parents, teachers, students—who feel that education is better for incorporating music in the curriculum will help themselves to this new model we devised. Together, we may enable children, especially educationally disadvantaged children, to tune in to more of their potential.

This book does not have formulas for creating young geniuses; nor is it a book of science. Rather, it tells a story, describes an educational process, and attempts to share some insights into the world of cognitive neuroscience. The reader will find possibly unfamiliar musical or scientific terms explained in the Glossary, at the end of the book.

—Peter Perret

ACKNOWLEDGMENTS

We would like to express our appreciation and thanks to the musicians who contributed so much to the evolution of this music residency project:

Robert Franz, Elizabeth "Lisa" Ransom, Bob Campbell, Dr. Eileen Young, Kristine Kohler-Hall, Steve Jones, Sonja Condit, Jon Julian, Cara Fish, Debi Reuter-Pivetta, James Kalyn, Kendall Wilson, Kathy Levy, Amber Ferenz, Carol Bernstorf, and Tim Papenbrock. Bravo!

Christine Griffith and Marie Wallace, both former staff members of the Winston-Salem Symphony, had the vision and the determination to help get the project up and running. The Winston-Salem Symphony Association's new support group, the Prelude Circle, headed by Patty Brown and Sue Henderson, gave extraordinary moral and financial support to the residency that came to be called the Bolton project. Dr. and Mrs. Malcolm Brown, and longtime Symphony patron Mrs. Gordon Hanes, made it possible for the Symphony to retain the services of Robert Franz as project coordinator. In the Winston-Salem/Forsyth County school system, Dr. Jane Pfefferkorn, Dr. Ann Shortt, Patricia Holiday, and Mary Norman were the early champions of this unorthodox method of incorporating music in the basic curriculum. At the Arts Based Elementary School, Mary Siebert's understanding and wholehearted support has been invaluable. Thanks mostly to the efforts and enthusiasm of these school administrators, the program is thriving ten years after its inception, still helping children learn and still adding to our knowledge of how music can help them do it.

Dr. Frances Rauscher was one of many scientists who have helped us try to understand what is happening in our musician residencies. She and Peter covered many paper tablecloths in restaurants with questions and ideas for future studies. At the Wake Forest University School of Medicine, Dr. Frank B. Wood and his colleagues in the neuropsychology section—Dr. Jonathan Burdette and Dr. Lynn Flowers—have been extremely generous in educating us about the brain, directing us to excellent sources of information, and reading portions of the manuscript for scientific accuracy. Any misinterpretation of scientific research that may have slipped through is ours, not theirs.

We are grateful to Jane Nevins and Dan Gordon at the Dana Press for helping us shape and focus our ideas, and to the scientific reviewers who helped us eliminate glaring errors from the manuscript.

We are delighted that Susette Sides offered her artistic talents for this book as soon as she heard about it. She brought her sketch pad to every lesson the woodwind quintet taught in the fall 2003 semester. Her charming drawings show, as words cannot, the warm relationship between the musicians and the children. More than that, her art stands in for the art that we can't put on the page—the music.

We appreciate the full support our families gave us during the three years in which we wrote this book. Love and thanks to Debra, Zachary, and Michael Perret; Tom Tomlinson; Emily and Michael Martine; Lucinda Fox; and John Mancini.

Janet is indebted to her Artist's Way friends for their loyalty and encouragement. Peter knows he could never have written this without the gifts he received from his neurosurgeon father, George Perret, and his concert pianist mother, Margaret Perret. He owes to them his early appreciation of the worlds of neuroscience and music.

Finally, we feel lucky to have had each other in the writing process. Without a doubt, our collaborative efforts have created new connections in each of our brains and enriched our lives. We hope this book will do the same for you.

—Peter Perret & Janet Fox
Winston-Salem, North Carolina
March 2004

CHAPTER 1

Fanfare

No person I have ever known has become a musician in hopes of improving his or her ability to think. Nor do I know of anybody whose love of music is based on anything other than the beauty of sound, the profound emotional and spiritual response, and the intellectual satisfaction the musical experience brings. The art of music is perhaps the most sublime human communication—ineffable, yet universally understood or felt. Even in its most unsophisticated and simple states, music is a powerful force that compels the emotions and often incites the body to motion. As such it stands alone, sovereign, without need of defense or justification.

Music has always been integral to education. Our ancestors knew this intuitively. Yet in our own time, music and education have parted ways in many school systems. As music came to be regarded as *art*—as opposed to a natural and instinctive human activity—it has been treated as a luxury rather than a necessity. My own bias makes me sure that its loss to general education is one important reason for the poor state of learning about which we complain year after year. This book is the story of how one school district and a woodwind quintet brought music back to school in a new and modern way and, by doing so, may have helped turn mediocre learning performance into high achievement.

In the spring of 1996, when the third graders at Bolton Elementary School in Winston-Salem, North Carolina, took the state-mandated tests in reading and arithmetic, they made a poor showing. Fewer than 40 percent of the children scored at or above grade level. Their mediocre performance was unsurprising. After all, this was a school population in which poverty, low IQs, and broken homes were more the rule than the exception.

One year later, the next crop of Bolton third graders took the same set of standard state tests. This time 85 percent scored *at or above* grade level in reading, and 89 percent were *at or above* grade level in math. Yet as far as the pupil population was concerned, the 1997 class was just like the 1996 class.

The children who excelled on the state tests in 1997 had had an extra element added to their instruction: an ensemble of classical musicians had taken up residence at Bolton in the 1994–95 school year, when these children were in first grade. Over a period of three years, the children's curriculum had been augmented by a flute, an oboe, a clarinet, a bassoon, and a French horn.

The quintet, made up of musicians from the Winston-Salem Symphony, was formed for the specific purpose of using music to improve the learning prospects of these at-risk students. From the spring of 1995 until 2002, the quintet visited classrooms at Bolton two or three times a week, for 30 minutes per visit, in residencies that were as short as seven weeks of the school year and as long as twelve weeks. The Bolton project, as it became known, has since moved to a charter school, where it can be scientifically observed and measured, but the musicians' approach has continued essentially as they developed it in the first year or two.

The musicians' lesson plans are integrated with the subject matter of the classroom teacher. The quintet members know the academic curriculum for each grade level. What they do in the classroom may clarify or extend a unit on arithmetic, poetry, teamwork, or any concept in the regular school curriculum. The quintet is not there to teach music, but to teach *through* music.

Each half-hour lesson begins when the musicians enter the classroom carrying their instruments, portable music stands, and sheet music. The quintet immediately launches into a short piece of music. Whichever musician has been chosen to lead that day's lesson then introduces the subject to the class. The other musicians, and the regular classroom teacher, fan out through the classroom. Within minutes, the children may be clapping out rhythms to understand how a half note is different from a quarter note; listening to a piece of music to learn about story elements like character, setting, conflict, and resolution; or standing up and waving their arms and stamping their feet to lock in their understanding of opposites like high and low, and bumpy and smooth.

The Bolton experience has attracted attention from many parts of the world. This different way of integrating music into the basic curriculum continues to be examined, refined, and extended to other schools and school districts.

Did musicians in the classroom directly affect mathematical proficiency? Did *A Little Night Music* help create a lot of bright readers? What does listening to music have to do with learning to learn? That's what I set out to discover, and that's what we continue to explore.

In my 25 years as music director and conductor of the Winston-Salem Symphony, some of my most rewarding work was the design and development of music programs for children. Whether we were playing for audiences of preschoolers or giving gifted young soloists an opportunity to perform with a professional orchestra, I saw over and over again how *live* music fills children with a joy that is hard to match. Through my work with young people, I witness the many ways that learning and listening to music enriches and ennobles their lives. The project I initiated at Bolton Elementary School built on those convictions, which I think most musicians share.

The quintet and I went to Bolton daring to hope that we could make a positive difference. And over those first three years, we had indications that we were helping. Teachers asked us what we were doing that had caused the improved attentiveness of the children. They told us that school attendance was better since the quintet had started visiting. Parental involvement in the school increased, and parents began planning fund-raising car washes to ensure the continuation of the program. The children's confidence and self-esteem seemed to grow. As word about the symphony musicians at the school spread, some children transferred to Bolton from other schools.

It wasn't until the test scores came down from Raleigh three years later, however, that we felt confident that we were making a difference. It happened to be the day of an orchestra rehearsal. As I read the test results, I could not keep my voice from breaking. And when I looked up, I saw my own emotions reflected in the eyes of the orchestra members. After three years of getting to know and care about those children, we were simply overwhelmed!

As laypeople, we cannot claim to know precisely what is happening in the brains of the children we teach. What we can do is relate in some detail what the musicians do in the classroom, and describe current scientific thinking that sheds light on why and how human beings are quintessentially musical beings. This is the story of how we use music to teach children to listen, and why we believe careful, active listening helps them learn to read and to reason.

CHAPTER 2

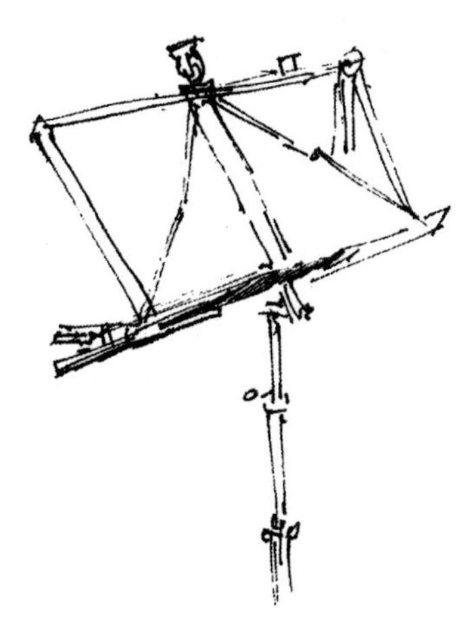

Lessons from the Pied Piper

Five strangers file into the classroom. Each is carrying a musical instrument. The group sets up chairs and music stands in a semicircle near the blackboard. The five adults sit down, raise their instruments to their mouths, and make eye contact with one another. Without a word, they play a piece of music from beginning to end.

The children watch and listen with rapt attention. At this first encounter, it's a safe bet that they know almost nothing about what they are seeing and hearing. Yet with this bright, concerted sound, the atmosphere is perceptibly transformed.

The music seems to pull the children into a state both calm and curious. Their bodies are relaxed; their faces are open and receptive.

As the musicians sound the last note of the opening piece and lower their instruments, the children sitting on the floor in front of them clap enthusiastically. Debra Reuter-Pivetta rises from the child-size chair she's been sitting on and greets the class. Onstage, as a soloist with symphony orchestras, she is the picture of sophisticated glamour, but this morning Debi has her hair pulled into a high ponytail, and she's wearing jeans.

"How many people do you see up here?" she asks. That's easy for these first graders.

"What to do you call a group of five people?" Debi asks. That's harder.

"We're a quintet, and we all play different instruments," Debi says. "My name is Debi, and the way I spell it is different from the usual way. I spell it D-E-B-I. I just like it that way."

Debi holds up her silver flute for all to see. She plays a few notes.

"The flute plays the highest notes of these five instruments. Do you know the little dog in the Taco Bell commercial?" With Debi's encouragement, the kids yip like Chihuahuas. "How about the big dog in the movie *Beethoven*? How does he bark?" Everybody makes deep Saint Bernard sounds.

"I'm the Chihuahua in the group. I play the high, squeaky notes," Debi says. She shows the class how her instrument comes apart into three pieces, and how she has to blow across the head joint to make a sound.

Next, Cara Fish rises and asks the class to participate in a scientific experiment. "Let's put on our lab coats," she says, miming the motions. "Now our safety goggles. Now put your hand on your throat and when I count to three, start humming...What do you feel?"

"Vibration," several of the kids pipe up.

"Right, vibration. For any sound, there has to be a vibration."

Cara goes on to explain that the quintet is a woodwind quintet, and they all make their sounds by blowing into the instruments. Their instruments are called woodwinds even though they aren't all made of wood. Debi's flute is made of metal, but some of the first flutes that were made were wood flutes.

Cara holds up her instrument, the oboe, and its reed, which she tells the class is made of two thin pieces of cane. The two pieces vibrate against each other to start the sound of the oboe.

"Do you think it will sound higher or lower than the flute?" she asks.

"Lower?" "Higher?" The children's answers seem tentative and unsure.

"It's lower," Cara says. "I'm a bigger dog. If Debi is a Chihuahua, I'm a cocker spaniel."

Cara blows into her reed, and the children laugh at the froggy sound it makes. Then she attaches it to the instrument and plays a little so the children can hear the sound of the oboe, contrasted to the flute.

Then Eileen Young holds up her clarinet. "What are some things about the clarinet that are the same as the oboe?"

The children raise their hands and observe that the two instruments are the same color, are the same shape, are made of the same kind of shiny wood, and have silvery keys.

"What are some things that are different?"

The kids notice a difference in size, but beyond that seem stymied.

"The clarinet has a reed, but it's a single reed. Want to hear my reed?" Eileen puffs out her cheeks and blows as hard as she can on the single piece of cane. No sound. The kids find this hilarious.

With a little prompting, the children figure out that the one-piece reed has nothing to vibrate against. Eileen shows them that the clarinet, unlike the oboe, has a mouthpiece.

"Want to hear my mouthpiece?" Again she puffs out her cheeks, blowing as hard as she can on the mouthpiece. Again not a sound.

Starting the clarinet's sound takes both the mouthpiece *and* the reed, fastened together. The kids don't need to be told that this is a bigger dog and will make a lower sound. Eileen plays the sound of the clarinet and asks the kids to look around and guess which instrument plays lowest of all. They point to Kendall Wilson and his big bassoon.

Kendall leads the class through the similarities and differences between his instrument and the others, and demonstrates the low, low notes he can play.

Then Bob Campbell talks about how the horn is similar to the other instruments in the quintet even though it looks much different and is a member of a different family from the woodwinds, the brass family. He shows them how he uses his lips to make all the air in the horn vibrate and puts his hand inside the wide bell to make the instrument blend better in a woodwind quintet.

The kids have been itching to ask questions, and after all the instruments have been introduced, the quintet recognizes them in turn. Why is that bassoon's reed so small? Why do you have a pencil on your horn? Do you ever play at football games? Where did you get that flute? How could you mail somebody a bassoon?

Twenty-five minutes have flown by, and after the questions there's just time for a closing piece of music.

"You may recognize this tune, but don't sing it out loud, because it's going to do some fancy things," Eileen says. "Sing on the inside, not the outside. Listen to the sounds of the different instruments, and see if you can tell which instrument starts the piece."

The members of the quintet sit down on their little chairs and launch into a lovely and elaborate arrangement of the folk tune "She'll Be Comin' Round the Mountain." Some of the children sway to the music. A few wiggle their fingers on the keys of imaginary instruments. All are intently focused on the musicians and the music, watching on the outside, singing on the inside, and remembering that it was the oboe that began the piece.

The time is a Monday morning in September 2003, and the place is the one-year-old Arts Based Elementary School in Winston-Salem. Essentially, this is the introductory lesson we've been presenting since we began at Bolton in 1995. Over the years the lesson has been adjusted for different grade levels, and it has been colored by the personalities of the musicians who were part of the ensemble at any given time and by the latest crop of learners.

In this classroom, like the classrooms at Bolton Elementary where the program was introduced, the children reflect the demographic makeup of the city—predominantly white, African-American, and Hispanic, and a smaller number of Asians. Although the quintet members don't know (and aren't interested in knowing), some of these children have already been diagnosed with learning disabilities or attention deficit hyperactivity disorder (ADHD); some are academically gifted. Although the quintet members don't know and aren't interested in finding out, some of these kids are the children of college professors, some are indigent, some are the offspring of professional artists, and some come from wealthy families.

What the musicians are reaching to and teaching to is something more fundamental than differences of race and income level and cultural conditioning. What they are communicating is a language both universal and ancient. What the children are doing when they respond to the sound of the music with their minds and their bodies is just what our human species has evolved to do.

People who have observed these musicians in the classroom over the years frequently comment that the effect of this experience of live music on the children is completely different from that of recorded music. Having real people playing instruments in front of their noses is riveting for almost all children. In contrast to the dopey, trancelike state that watching videos and television often induces in young children, the musicians evoke wide-awake curiosity and interest.

Sergiù Celibidache, the great Romanian conductor, who was one of my most influential teachers and mentors, used to say that listening to recorded music is like kissing over the telephone. There's a place for recordings, but I don't believe any CD or DVD could create the magnetic effect in the classroom that our live musicians consistently produce.

Our modern-day Pied Pipers are carrying on a tradition that is rooted in prehistoric time. Recent archaeological digs have unearthed musical instruments in Ukraine, made from the bones of a woolly mammoth, dating to 18,000 B.C., and an 8,700-year-old bone flute in central China that can produce a seven-tone scale. These and similar finds provide a record of musical instruments and musical activity across all human settlements, going back at least 50,000 years. It's likely that music is even older than that, dating from our evolution from quadrupeds to upright walkers. From that point on, our ancestors would have been able to use their vocal equipment to hum and sing. The soft humming sounds mothers make around their babies and the sounds that sometimes accompany mating rituals may well be the predecessors of spoken language.[1]

In addition to the music of intimacy, our hunter-gatherer ancestors also used musical sound to communicate over long distances. Some tribal peoples still use horns and drums made from animal parts for this purpose. Over time, music became a rhythmic and chanted accompaniment to rituals and routine tasks. Some scholars believe that prehistoric tribes did not play music for its own sake, but always as an integral part of some other activity.

Of course, music still plays an enormous social role as accompaniment. Religious ceremonies, military campaigns, and sports events depend on music to unify people and bring them to the desired emotional state. Work songs, drinking songs, dance rhythms, lullabies—much of the world's folk music provides the beat for ordinary human activities. And it would be hard to even imagine a movie or television show without its driving musical score. Music, as we see at football games, piano bars, graduation ceremonies, bar mitzvahs, rock concerts, and countless other gatherings, contributes to a sense of togetherness.

As Isabelle Peretz, one of the most prominent researchers into the biology of music, writes, "We sense that music is much more than a pleasure technology or mere entertainment. Yet, the idea that our responses to music might be instinctual has only gained credence in recent years."[2]

Dr. Peretz, a professor of cognitive psychology at the University of Montreal, writes that human beings appear to be born with the predisposition to process music, and that the ability develops simply by being exposed to music. She believes that the mother's singing is the initial trigger.

More evidence that we are all born musical comes from the work of Beth Bolton (no relationship to Bolton Elementary), a music educator and researcher and curriculum director of the Early Childhood Music Foundations Program at Temple University Community College. In her studies, babies were exposed to wordless songs, in a variety of modes, sung to them in comfortable settings. Invariably the babies responded by cooing or vocalizing a sound on either the tonic (do, or 1, of the scale) or the dominant (sol, or 5, of the scale), regardless of the key or mode of the song.[3]

Other studies have indicated that newborns have the ability to remember music from one day to the next.[4] In the words of Dr. Sandra Trehub, a well-known researcher in the field of early childhood cognitive development at the University of Toronto, "It is clear that infants do not begin life with a blank musical slate. Instead, they are predisposed to attend to the melodic contour and rhythmic patterning of sound sequences, whether music or speech."[5]

Another researcher, Dr. Jenny Saffran, at the University of Wisconsin in Madison, concludes, "...it is evident that even the youngest listeners are already accomplished musicians in their own right."[6]

In all cultures, mothers naturally and spontaneously sing to their babies. The way mothers and other people talk to babies, sometimes called motherese, likewise has a singsong quality. Stephen Malloch, of the University of Western Sydney, MacArthur, Australia, has shown that newborn babies also respond to their mothers' voices in a musical way. His analysis of the structure of vocalizations of mother-and-baby pairs found that the infants' responses were purposeful. The mother's pitch also changed to resemble her baby's vocalizations, apparently signaling to the baby that the mother had received the baby's message. These "duets" are thought to be important in the bonding of mothers and babies.[7]

Several scientists and anthropologists have suggested that language may have evolved from this motherese. One of them is anthropologist Dean Falk of Florida State University, who bases her belief on a study she has done in collaboration with German scientists. She believes that tone of voice opened the way to understanding, and that the process may have started with the way in which mothers communicate with their babies.[8] "In evolution, motherese was preparatory to language," she says.

Such studies suggest that human beings have evolved to be musical beings, and it's likely that something about music has helped our species survive. As Isabelle Peretz notes, Darwin himself proposed that music helps us attract sexual partners. Another theory is that music helped our species survive because it helps members of the tribe bond. In fact, music still serves both these purposes today, and what advances courtship and cements community is bound to be potent stuff.

Dr. Peretz believes that the universal appeal of music is actually an adaptive response of the human organism, and that our responses to music are involuntary reflexes. Unless we have a particular neural impairment, we can't help responding to music—physiologically and emotionally.

That might explain why the Pied Piper of Hamelin was able to lure the children of the village away from home and family with nothing more than his own breath blown through a reed. And it may explain why our symphony musicians in the classroom can so predictably capture the full attention of each group of children. Music touches children with the force of instinct.

CHAPTER 3

Overture

The idea of a link between music and learning is at least as old as Plato, and as new as the educational toys currently on display at the mall. Yet for centuries the music-brain connection has been a matter of conjecture and intuition, not hard evidence or scientific proof.

Take, for example, the belief that hearing the music of Mozart will raise a baby's intelligence. Many American newborns own the beginnings of a classical compact disc collection before they leave the hospital, courtesy of state legislatures, manufacturers of baby food, and trendy well-wishers. Babies of ambitious and upwardly mobile parents imbibe Baroque classics along with breast milk and take Kindermusik classes between diaper changes—all thanks to a letter to the editor of a scientific journal that captured the popular imagination.[9]

No scientific evidence will ever show that exposure to classical music in infancy will raise intelligence. Babies' brains don't work that way. Still, it is not surprising that the public has been so ready to embrace the idea. Most of us have never needed scientific proof that music can exert a profound effect on how we feel. Many would say that music mirrors their moods and emotions with a precision that words can never match. We feel an intimate connection to music that borders on the uncanny. If music can make our pulse race, and our tears flow, why shouldn't it be able to alter the way our brains function?

For most of the history of Western civilization, the importance of music in the general education of children was unquestioned. Plato went so far as to prescribe particular types of musical training for future warriors and leaders. He suggests, for example, that the learning of certain modes or scales that express sorrow are unfit for the training of the "Philosopher King."[10] In his convictions about how music builds character, the ancient Greek philosopher seems to be echoing the prevailing wisdom of his time.

In our time, music and the other arts have come to be regarded—at least by many political leaders—as frills or luxuries in the school curriculum. Music classes are often the first to go when budget cuts must be made. Our forebears may have been wiser about the fundamental role of music in the development of individuals and civilizations. And music may be even more basic than Plato imagined. The relationship between music and the mind is today not just a subject for philosophical musing and conjecture but for scientific research. Advances in neuroscience and technology will soon make it possible to discover how music affects and perhaps even alters the brain, and to examine the interplay between music and intelligence.

I should probably mention here that science, particularly brain science, is one of my lifelong interests. My father taught neurosurgery, and I had seriously considered a career as a scientist. I should also mention that I am the father of three, and that as I write this my youngest child is nine years old. So it is not surprising that anything that involves music, science, and the education of children is guaranteed to pique my interest.

I had long suspected and hoped that there might be a way to find out whether music could somehow enhance academic performance. One day in 1992 a news item on National Public Radio riveted my attention. It described a study in which young children in day-care centers were taught to play music keyboards. These children scored higher than their peers on tests of a particular kind of intelligence—spatial-temporal reasoning. Spatial-temporal reasoning, as the term suggests, involves thinking that uses spatial representations in time, such as sequences and visualizations of objects being rotated in space. It is basic to mathematics and to abstract thinking in general. I'd always had the idea that music and intelligence were in some way connected, but this was the first scientific evidence I had encountered that seemed to indicate that learning to play an instrument at a very young age might make a person smarter.

I was so intrigued by that little news item on NPR that I sent away for a transcript of the broadcast. When I received it, I wanted to know more, so I telephoned the University of California at Irvine and ordered an advance copy of the study. I learned that one way to test spatial-temporal reasoning ability in a three-year-old is to present a picture, cut into several pieces, of a familiar animal and ask the child to put the pieces together. The children who had learned to read and play music exhibited more speed and accuracy at this task than children in control groups.[11]

The idea that playing music might develop spatial-temporal thinking ability was intriguing. I could well believe that learning notes, reading ahead, memorizing fingerings, hearing tones, figuring out rhythms—and doing all those things at once—might stretch the brain in some way. It seemed likely that the skills involved in playing an instrument might carry over to other complex mental tasks.

I began to wonder whether it was necessary to physically perform all those musical actions oneself to improve mental performance, or whether simply being exposed to a certain amount of live music might also be able to energize the brain. I remembered reading, in the 1970s, about plants whose growth was affected by exposure to different kinds of music. I recalled that famous line: "Music hath charms to soothe the savage breast,/To soften rocks, or bend a knotted oak."[12] I was certainly open to the idea that music has powers we have yet to discover.

Then and there, in 1992, I decided to find a way to test this.

FROM A QUESTION TO A QUINTET

As the music director of the Winston-Salem Symphony, I had overall responsibility for the Symphony's education programs, which were long-standing and reached all children in the local public school system. String, wind, brass, and percussion ensembles went out to the elementary schools, where they performed and presented information about their instruments and the music. The ensemble visits culminated in a full orchestra concert in a large auditorium, to which classes from all the schools were bused. For many children, of course, these close encounters with the symphony are their first, and often their only, exposure to classical music and to the live performance of music. For some children every year, they are life changing, opening up a new world and setting some boys and girls on the path to becoming musicians themselves.

Valuable as this type of traditional education program is, my feeling was that these brief encounters would not be enough to actually have an impact on the learning skills of the children. I felt pretty sure that it was the act of learning to play an instrument that somehow boosted mental ability. At the same time, I thought it would be wonderful if mere exposure could have a similar effect. Realistically, we're probably far away from the day when all primary school children could own or rent instruments and have lessons with a music teacher. But making sure that all children hear live music as part of their schooling seems quite practicable.

Although symphony orchestras are not particularly qualified in scientific areas, I hoped to make a serious effort to find out how music might advance academic goals. One element that seemed important to me was that the exposure to music be as regular and frequent as possible. While I was thinking about who could help turn an interesting hypothesis into an effective in-school program, I tried to keep up with the research world and the ongoing experiments of the psychologists at Irvine, as well as those of other neuroscientists, particularly at Harvard.

One of the first people with whom I talked about my ideas was Dr. Jane Pfefferkorn, who at the time was program manager for school-community cultural arts in the Winston-Salem/Forsyth County school system. Jane had responsibility for special arts programs in the public schools at all levels, bringing artists from Winston-Salem's many cultural organizations into the classrooms in all sorts of enlivening and innovative ways.

I talked to Jane about my initial idea of having musicians in residence in one or more elementary schools. What I envisioned was musicians rehearsing in the halls and other common spaces in the school, where children would hear music as part of their daily life and also have a chance to observe the way members of a musical ensemble work. I thought there was something kids could learn from the way musicians practice difficult passages over and over, how they communicate with one another through nonverbal gestures, and how they resolve differences and make decisions.

I think Jane immediately realized that this notion would not fly with most principals. Having a group of musicians hanging around in the halls might well sound more like a distraction than a way to raise test scores. But she said there were two elementary school principals who were likely to be receptive to taking this completely new idea and developing it. One was Dr. Ann Shortt, who had worked as an administrator in the central office of the city-county school system, and who had recently volunteered to become principal at Bolton Elementary School. Few people would have seen this as a plum assignment, but Dr. Shortt was determined to make a positive difference in the lives of these children.

I began meeting frequently with the people from the school system to talk about how we could turn my ideas into an effective program. Our meetings soon grew to include some members of the Symphony staff, along with a potential corporate sponsor. Early on we foresaw the need for a way to link the musicians' efforts with the lessons taught by the classroom teachers. We quickly began to talk about having a Music Master, a person who could make connections between music, the needs of the classroom teachers, and other areas of the children's lives. From the field of possible candidates, we almost immediately identified Robert Franz as the consummate connector, the ideal Music Master.

Robert was suited to this largely undefined job by both experience and temperament. At 25, he was an accomplished musician, with his bachelor's degree in oboe performance and his master's degree in conducting, both from the North Carolina School of the Arts. It was there that I first knew him, when he was a young oboe student and I was head of the orchestra program. As a student, Robert had been a founding member of three musical groups, including the Carolina Chamber Symphony, of which he was the music director and conductor. He had already demonstrated quite remarkable gifts at starting new ventures, promoting them, and garnering support for them.

When I called Robert in the summer of 1993 to see if he might be interested in the Music Master role, he had just finished a nine-month stint in Liberal, Kansas, working in a Rural Residency Chamber Music Initiative. This program, sponsored by the National Endowment for the Arts and administered by Chamber Music America, had as a major component the development of educational music programs for people of all ages. Robert was in Winston-Salem for the summer, and he was planning to do another such residency in the coming year, this time in Fitzgerald, Georgia. I urged him to go ahead with that and gain more valuable experience, as it appeared that we would not have our project fleshed out and funded for about a year. He completed the residency in Georgia, and that additional relevant experience paid off handsomely.

Robert's personal qualities were every bit as important as his professional credentials. The two words I think best describe him are *fearless* and *childlike*. Once he decides to undertake a project, he does not get bogged down in worries and worst-case scenarios. He enjoys taking risks and is not deterred by the obstacles that accompany the plowing of new ground. Robert is childlike in his enthusiasm. He appears to take joy in almost everything he is involved in, and his enthusiasm is contagious.

He connects easily with people of all ages, and his personal interactions have a playful, light-hearted quality.

Our informal design team drew up what we thought would be an ideal way to set up our program. We envisioned having two ensembles, a string quintet and a wind quintet, working in two schools. We wanted to have the musicians at the schools all year, for two or three hours a day—rehearsing, going into the classrooms individually and in groups to incorporate music into the academic curriculum, and offering private music lessons. And we intended to make the state-prescribed tests at the end of third grade the benchmark of the impact of the music residency.

Our readiness to be held accountable, along with the fact that our effort had been inspired by scientific research, pulled a lot of enthusiasm and energy to the project in its conceptual stage. As the project took shape, we were able to translate that enthusiasm into financial support that would ensure its continuation for at least three years, long enough to start with a crop of first graders and see them through those third-grade tests.

The North Carolina Arts Council provided a grant to implement the project at one school, but our other prospective funding source had to withdraw because of policy changes at the corporate level. So when the design process was complete, we had only one school, Bolton Elementary; one musical ensemble, a woodwind quintet; and a strong focus on the whole quintet working in the classrooms with whole classes. Our quintet did rehearse in the halls at times, but that never had the impact I once imagined it would. Individual music lessons at the school from members of the quintet never materialized. Nor were we ever able to afford to have musicians in residence for more than about a quarter of the school year. But despite these apparent shortfalls, the project flourished and never left us with the sense that anything was missing.

What happened with our fledging idea was something like what happens with every baby's brain. In the beginning the potential is virtually unlimited, but not all the potential is realized. Early on, some possible hookups fall by the wayside. Other connections survive and are strengthened by use.

And so it was with the Bolton project. We didn't receive the funding needed for additional schools, musicians, or hours of class time. Some potential thus fell by the wayside. In moving forward with our adventure, however, we didn't stop to mourn what might have been. We were buoyed up by the thrill of being pioneers, of inventing new ways to illuminate the elementary school curriculum and perhaps show that music nourishes the mind.

MEET THE PLAYERS

The School

At Bolton Elementary School in 1994, the average IQ was 92.[13] A full 70 percent of pupils qualified for free or reduced-price lunches because of household poverty. A high proportion lived with a single parent, a relative other than a parent, or in foster care. A fair number were transient or homeless. As previously noted, more than 60 percent performed below grade level on state achievement tests.[14]

Yet under the leadership of Ann Shortt, the school had recently become an unusually warm and inviting place. A visitor would have been struck by a colorful mural painted on the outside wall, groups of teddy bears arranged on sofas and chairs in the lobby, and the sounds of Vivaldi coming from the intercom. (Typical of her openness to any ideas that might help the children, Dr. Shortt had introduced classical music into the environment well before the Bolton project began, just days after a brief conversation I had had with her.) The school was a home and refuge to many children and their parents.

The daunting demographics of the school, combined with its genuine desire to create change, made Bolton an ideal environment for trying out what may have looked like a far-fetched proposition. We went into the school with some educational models in mind and with the intention of taking our cues about what the children needed to learn from the classroom teachers. We went in with the understanding that we would not be allowed to devise and administer any tests of our own to see how the project was working. The only objective indication of the possible connection between music and academic performance would come at the end of the third year, when our initial group of first graders took the third-grade standardized tests to measure math and reading competency.

In her tenure as principal at Bolton, Dr. Shortt did a great deal to show how disadvantages can be transcended and how the culture of a school can be transformed. She believes the music residency was a major player in that transformation. Not all of her innovations were continued after she moved to Alaska, where she is now superintendent of the Fairbanks school system. Well before the third-grade test scores were in, Dr. Shortt was talking enthusiastically to her fellow principals about the music residency. One of them, Patricia Holiday, was able to implement a similar program, using the same woodwind quintet, at her middle school.

The Musicians

While my first concern was finding ways to benefit the children, I also saw the project as an excellent opportunity for the musicians. Classical musicians, who have invested many years in their training and many thousands of dollars in their instruments, typically struggle to earn a decent living. The Bolton project would provide some additional income to an ensemble of symphony musicians. It would also offer them a couple of other benefits of significant value: the chance to acquire teaching skills that would give them more control of their lives and make them more employable, and the time and a place to practice playing chamber music.

The ongoing development of the artistic quality of the orchestra was also on my mind. The original members of the Bolton quintet were the second-chair players in each of their sections. Whereas the principal players already participated in the well-known Clarion Wind Quintet,[15] this school program would give these second-chair players a durable ensemble and all the improved accuracy and balance that come from regular ensemble playing.

Of course, not all orchestral players would be good primary-grade teachers, or would even be interested in trying. But it hasn't been difficult to find members of the symphony who relate well to children, who can translate the objectives of the regular classroom teachers into lessons that can be learned through the sounds of music, and who are extremely creative in developing a program that uses musical knowledge and musical instruments to teach arithmetic, reading, social skills, and many other elements of the curriculum of the primary grades.

Robert Franz was the oboe player in the quintet in the project's first two years, and coordinator for the first three years, after which he moved to Louisville, Kentucky. As associate conductor of the Louisville Orchestra, music director of the Louisville Youth Orchestra, and music director of the Mansfield (Ohio) Symphony, he continues to teach and expand upon what he learned from the Bolton project. Robert's style in the classroom was easy, direct, and informal. In the first year or two, the other quintet members looked to him for strong direction, but he was comfortable with not jumping to solutions too quickly. Serious fishing for sound and effective teaching strategies was his initial approach.

Although we did not think much about it at the time, differences in temperaments, teaching styles, and backgrounds of the musicians have proved to be a great strength of the residency project. Over the years, there has been a gradual turnover in the group.

When we talk about the Bolton quintet, then, we are talking about different musicians at different times and at different schools. So many exciting pilot programs in schools are dependent on one charismatic individual. When that individual leaves, the results typically take a nosedive, or the program is discontinued. We had a charismatic leader in Robert, yet the residency project has continued to thrive many years after his departure. With different sets of musician-teachers, in other schools, and at other grade levels, the curriculum and methods developed by the original quintet have been maintained. While the players have changed, the score has changed very little.

The Brain

When the Bolton project began, the human brain was the mystery player. Since then, our knowledge of this most complex thing on earth has expanded enormously as findings from the world of science have made their way into the mainstream. Even more recently, research on the music-brain connection has gotten under way. And the news is good. We now know that the brain is much more plastic than had been imagined, ever ready to create new cells and form new connections. Far from being a blank slate, our brains are hardwired, or preprogrammed, for many of life's vital functions. Far from being fully formed at birth, though, our brains require interaction with the external environment to mature. Far from being highly compartmentalized, our brains are highly flexible. If one part of the brain is injured, other parts of the brain may pick up its functions. If one pathway is blocked, the brain can create new pathways and other connections.

Because of the amazing plasticity of our brains, our human potential is far greater than we ever knew it to be. That implies that we should never give up on the ability of children—or anyone—to learn. Even children with some obvious strikes against them are nowhere near as limited as was once believed. Intelligence and talent are not fixed and static qualities; they are subject to change at every moment.

CHAPTER 4

Getting to Know All About You

"Who remembers my name?" the flute player asks after the quintet finishes its opening number, an Allegro by Mozart.

It's only the quintet's second visit to this first-grade classroom, but most of the children remember Debi's name.

"Who remembers the name of my instrument?" Debi asks.

"The flute, the flute," the kids call out confidently.

"Hands, please," Debi says, raising her own hand to show the children what to do to be recognized when they want to answer or ask a question.

"Do I play high notes or low notes?"

The kids wave their hands in the air, but this time they wait for Debi to call on one student, who correctly answers, "High notes."

"That's right, I play high notes, because my instrument is small.

"I've asked you some questions. Now you can ask me some questions. What are some question words? *What* is a good question word."

"How about *how*?"

"Yes, *how* is a question word."

As Debi interacts with the children, Eileen writes "What" and "How" on the board.

"What kind of animals do you have?" one child wants to know.

"I have a gray cat. His name is Oscar," says Debi.

"Where did you get that flute?" another child asks.

"*Where* is a good question word. I went to a flute convention. You can get them there."

"What year did you get it?"

"I bought this flute in 1990."

"When did you start playing the flute?"

"*When* is another question word. I started playing in fifth grade, in school."

"My sister has a flute."

"That isn't a question, is it? Do you have a question?"

"Debi, where are you from?"

"I was born in Wisconsin. They make lots of cheese there."

"Where is Wisconsin?"

"It's north and a little bit west of here."

"Do you have a husband?"

"Yes, I do have a husband. He's Italian and he's a musician, too."

"Do you live in Italy?"

"No, we live in Winston-Salem."

"Do you eat a lot of spaghetti?"

In the next 20 minutes the children learn that Cara rhymes with Sara and begins with a "k" sound. Some children remember that the name of Cara's instrument sounds something like "elbow." They learn that Cara's oboe comes from France, and that she grew up in Oklahoma. She also owns a cat, named Boxer because he punches his paws out like a prizefighter when he is petted. They find out that Eileen's clarinet has a bell on the end, where the flute has a foot joint. They learn that the quintet first started playing together in 1995, before the children were born. They discover that Kendall has one kind of key in his pocket and another kind of key on his bassoon. They learn that Kendall bought his instrument from a bassoon repairman in Indiana but that the instrument is over 70 years old and was made in Germany. They find out what a composer is, and that Mozart was a composer, and that some of the music he wrote was dance music.

This lesson is specifically about learning to distinguish questions from statements, which is not always easy for first graders. It also teaches children how to find out what they want to know. No questions are out of order here. If the question is important to the child, it's important. No question gets short shrift. Even when a simple yes or no would answer the query, members of the quintet invariably provide additional information. This keeps the dialogue flowing and the interest level high.

In the second and third grades, the information-gathering process goes further. The classes are divided into smaller groups and each group "adopts" one member of the quintet and is charged with learning as much as possible about that musician and that instrument. With some help from the classroom teacher before the musicians' visit, the children prepare a list of interview questions. As clarinet specialists, for example, they might ask what the clarinet is made of, when it was first invented or used, how much it costs, how much it weighs, how it comes apart, and how its sound is produced. As part of "adopting" Eileen, they might ask her why she chose the clarinet, how long she has been playing it, what other groups she plays in, whether she teaches people to play the instrument, what countries she has traveled to as a musician, and much more. Eileen may bring in other clarinets so that the children can see small differences between instruments that are essentially the same. The children may examine and measure the clarinet. They will also trace Eileen's silhouette as she lies on the floor and then color in her form and mount their life-size Eileen in the classroom.

With all this information, each group then prepares a written description of its musician and instrument. With help from the classroom teacher, the children work to shape the information into cohesive, effective presentations. They then present their musicians and instruments to their peers.

Like all the lessons from the quintet, the "getting acquainted" exercises work simultaneously on several levels and fronts. Each lesson repeats and builds on what the children learned in a previous lesson. (Corporate trainers in recent years have come to understand that retention of new knowledge depends on this kind of spaced repetition of the information—something musicians know from their musical training. Reviewing and repeating information before adding new information or techniques is standard in musical instruction.) The children's natural curiosity is directed toward exploring a topic deeply. Their questions are met with rich information that adds to their store of knowledge of physics, history, geography, and many other subjects. Their collaborative skills are developed by working in small groups or teams. Their ability to retain and remember what they have learned is enhanced by teaching it to others. Their self-confidence is strengthened by becoming "specialists" in a given subject and by standing up and speaking in front of the class. And the good humor, friendliness, and openness of the musicians build an emotional bond with the students.

Our musician-teachers seem to have understood intuitively that they had to arouse the curiosity and interest of the children before any learning could take place. Early on, the quintet members observed that some classes they visited were abuzz with this kind of energy, while other classes seemed dull and dispirited. In fact, that is how the quintet was able to tell what the regular classroom teachers were like. If the class had a good, inspiring teacher, the children were alert and ready to learn the minute the quintet walked in the door. If the regular teacher was the kind who squelches curiosity, the musicians had to warm the children up before they could start the lesson.

The members of the original quintet quickly realized that lively interaction was the way to get and keep the attention of the children. Playing a piece of music at the beginning of the class dependably commands the initial focus. When they are not actually playing their instruments, the musicians rely on questions to keep the interest level high. The children are encouraged to ask questions, and the musicians ask questions of the children from their very first visit: "What do you think this instrument is made of?" "Will the clarinet play higher notes or lower notes than the oboe?" "Which instrument first played the melody in that piece of music?" "Does this music make you feel happy or sad?" "How is the flute different from the bassoon?" "How is the flute *like* the bassoon?"

The kinds of questions the musicians pose are about things that the children can be expected to know, with some observation and thought. If a child's answer is incorrect, the musicians are likely to ask another question rather than tell the child she is wrong. If the child says the clarinet will play higher notes than the flute, the musician might say, "Take a look at these instruments. Which is bigger? Do bigger instruments sound higher or lower?" If the question is subjective, the musicians always indicate that any answer can be "right." They may say, "Yes, this music makes some people feel happy. It kind of makes me feel sad, but it makes some people feel happy. It can make you feel happy or sad."

The musician-teachers depend heavily on drawing answers out of the children rather than simply piling information on them. This style of teaching not only engages the minds of the kids in a fruitful way; it also conveys respect. It implicitly tells the children that they know things, that they can figure things out, and that their observations have value.

This effective teaching style is, of course, as old as Socrates. In fact, the method of leading students through a series of easily answered questions takes its name from the ancient Greek philosopher. In the Socratic method, the teacher knows what he wants the students to learn, but instead of simply telling them, he draws the answers and conclusions out of them. He allows them to discover the truth through their own thought processes. This is the most fundamental meaning of education, from the Latin word *educere*: to lead out from.

So one of the very first things our musicians in residence do is evoke the natural curiosity of children, using the sight and sounds of the musical instruments to do so. A new thing enters the children's world—a musical ensemble in the classroom—and they are encouraged and empowered to find out about this phenomenon by the musicians' Socratic teaching skills.

A WINDOW ON THE BRAIN

One method doctors and scientists use to find out about the body and its functions involves taking pictures of it, now called imaging. Old ways of seeing the brain, such as computed tomography (CT) scans, were static, like snapshots, though they sometimes had refinements, such as the ability to picture the brain in a series of progressive slices, or even in three dimensions. Another way of learning about the brain was through electroencephalography (EEG), whereby sensitive electrodes are placed on the scalp to record electrical activity at the surface of the brain. Newer imaging techniques, including functional magnetic resonance imaging (fMRI), positron-emission tomography (PET), and magnetoencephalography (MEG), allow scientists to see which parts of the brain are being used *while* the subjects perform various tasks.

To test a hypothesis, research scientists need to perform experiments that can be repeated under controlled conditions. Studying math problems and pondering moves in a chess game have too much variability to be of much use in scientific research. But a recording of a Mozart sonata or a Haydn symphony can be played over and over again. Thus, for example, the researchers at the University of California at Irvine reasoned that music might provide a window into the workings of the brain.[16]

Other studies, while more controversial, also provide food for thought:

- In one famous study, college students listened to music before taking a battery of IQ tests. Those who listened to a Mozart sonata scored higher than those who listened to rock-and-roll oldies, relaxation tapes, or contemporary minimalist music, but the effect was short-lived. This study is the source of the term Mozart effect, which captured the public imagination but which proved nothing about the effect of classical music on the brains of infants or anyone else.[17]

  Other researchers have been unable to duplicate the results of this study and have questioned what actually accounted for the higher performance in the tests. For example, the music used, Mozart's K.448, is a quick and brilliant piece in a cheerful major key. We know that people generally function better when in a happy mood, which would be the probable effect of listening to this music. In addition, the quick pace and high energy of the music would be likely to elicit a state of heightened arousal in the subjects, which might also account for the better performance.

- Then there is the intriguing evidence that parts of the brain are larger in professional musicians than in people who work in other occupations. Specifically, the corpus callosum of musicians is enlarged. This is the bundle of nerve fibers that connects the right

> hemisphere of the brain to the left hemisphere and vice versa. Having an outsize corpus callosum is not only a matter of having more fibers but of having more *insulated* fibers, which are the parts of neurons that transmit electrochemical messages. An interesting wrinkle is that musicians who started playing their instrument before age seven have the largest corpus callosum of all.[18]

But which came first, the violin virtuoso or the big brain? Recent studies indicate that the increased size of the corpus callosum is the *result* of learning to play a musical instrument.[19] This is not really surprising, because playing a musical instrument involves the use of many parts of the body, and therefore the brain, all at once. Some of those body parts are fingers, hands, arms, and feet, for pedals. The sense of sight helps us find the notes on the piano. The sense of hearing enables us to correct and adjust intonation. Higher cortical functions (thinking that does not involve direct sensory or motor input) tell us when to play softly, when to speed up, and how long to wait between sections of the piece.

Playing music is one of the most complex—if not *the* most complex—activities a person can undertake. Learning to listen carefully to the elements that make up music is less complex, yet it is clear that our musicians in the classroom stimulate questioning and problem solving, primarily and effectively through the *sounds* of their instruments. Getting to know a great deal about the particular sounds of music is a salient feature of our residency programs, and something that distinguishes them from most other arts-in-the-curriculum initiatives.

In the early years of the musician residency, I was skeptical that simply listening to music could have an effect on children's academic performance. Could learning to tell the difference between an oboe and a clarinet—with the eyes closed—have anything to do with reading, writing, and arithmetic? Would the ability to recognize a passage of music as staccato or legato have any measurable impact on general learning? I had my doubts.

CHAPTER 5

Building Bridges

"When we started the Bolton project, I was a little bit arrogant—until the first meeting with the classroom teachers," Robert Franz remembers. "At that meeting the teachers said, 'Oh great, we can use you at rest time.' I realized I had no language to talk to them about what we were trying to do."

We all knew that getting the support of the teachers would be essential. If the quintet was going to teach and reinforce the basic academic curriculum, the musicians would need a good understanding of what the classroom teachers were teaching from week to week, and where they could use help in getting particular concepts and skills across.

Above all, we didn't want the teachers to have the perception that the symphony musicians were wasting valuable instructional time. We were well aware of the pressure they faced to get through the state-prescribed curriculum and prepare their students for the state tests. The heavy emphasis on testing, including evaluating schools and teachers on how their pupils perform on standardized tests, has a long history in North Carolina, and testing mania was well under way by 1994. We wanted the teachers to understand that we were there to help, not to take time away from essentials, and certainly not to provide mild relief at rest time.

Robert, who was the first coordinator of the Bolton project and who had major responsibility for developing the curriculum, had some experience in figuring out how to make a lecture/demonstration in music reinforce a study unit in a physics, French, or Renaissance history course. And like most symphony organizations, ours was adept at providing themed in-school programs to enrich some aspect of the curriculum. The challenge at Bolton was the young age of the children and the lengthy intervention, a far cry from the usual one-shot program. When Robert first met with the classroom teachers he saw that it would not be an easy sell.

Robert came away from the meeting with the sense that educators and musicians use different vocabularies to talk about concepts and objectives that are substantially similar. He began looking for a common platform on which the teachers and the quintet members could understand each other's worlds. In the symphony office, he came across a book I had brought in for other staff members to read—Howard Gardner's *Frames of Mind: The Theory of Multiple Intelligences*.[20] The book describes many types of intelligence, not only the intelligence measured on IQ and achievement tests. Gardner describes linguistic, musical, logical-mathematical, spatial, bodily-kinesthetic, interpersonal, and intrapersonal intelligences. The Bolton faculty members were already familiar with the theory, and Robert quickly saw that musicians use all seven types of intelligence in varying degrees both to learn and to play music.

Reading music is a linguistic task, for example, while *rhythm* involves logical-mathematical intelligence. Playing an instrument draws on bodily-kinesthetic ability; and ensemble playing is both a spatial and an interpersonal challenge. For these skills to add up to something pleasing and meaningful requires musical intelligence.

The teachers at Bolton were aware of, and were incorporating into their teaching, other methods based on preferred modes of learning, or learning styles. The VAK model—visual, aural, and kinesthetic—also resonated deeply with the musicians, who use their eyes, ears, and body movement to learn, reinforce, and remember every piece of music they play.

If you think about how you learn things as an adult, you will probably see that you depend on some combination of visual, aural, and kinesthetic modes. As you listen to a lecture (auditory), you may find that you learn and remember best if you also take notes (visual and kinesthetic). As you read a book, you may highlight key points or even "photographically" memorize the position of text and illustrations on a page. When you study for a test, you may find that it helps to outline subjects, or draw maps that connect various parts of the study subject, or orally teach the material to someone else. You probably learned the alphabet and the multiplication tables by rote, chanting them aloud like an epic poem or even singing them. The VAK model can be applied to almost any kind of learning, so it formed another bridge between musicians and primary-grade teachers.

Weeks before the musicians were scheduled to go into the classrooms for the first time, the quintet and the teachers met for a daylong workshop. The musicians explained and demonstrated how they used all seven types of intelligence Gardner described, and asked the teachers how they drew on the different kinds of intelligence in the classroom. Using a matrix, musicians and teachers plotted which types of intelligence were important for which subjects, and how musicians used the same types of intelligence for similar or analogous musical tasks. The upshot of the day was agreement that music could be used to help teach almost anything in the curriculum, from vocabulary to science, from multiplication to social studies. The necessary initial acceptance from the faculty was achieved at that fruitful meeting, and as time went by a deeper trust developed, so that teachers didn't hesitate to ask the musicians for help in explaining and demonstrating difficult lessons and concepts.

At this first workshop, the musicians gained considerable insight into the needs and perceptions of teachers. Since then, the preprogram workshops have become shorter, and they can now be done in a couple of hours. Whenever the quintet goes into a new school, or will be working with a new group of classroom teachers, however, we see this kind of preparatory session as crucial to the success of the program.

The classroom teacher can be a real factor in how much the children benefit from the music program. Ideally, he or she participates in the classes the musicians teach, observes their methods and learns ways to help children who need extra help, and incorporates vocabulary and concepts from the quintet's lessons into classroom activities. In the real world, now and then teachers have used the quintet visits as an opportunity to take a break, grade papers, or talk on the phone. Distancing themselves in these ways, though, may send a message to the children that what the musicians are teaching is not part of the regular course of study or is not important, as far as the teacher is concerned. Whether the teacher decides to be a bridge or a barrier, however, the musicians have consistently been able to engage the children's attention, and to teach them.

Although putting professional musicians in an elementary school for weeks at a time was a new idea, incorporating the arts in the academic curriculum was not. Schools across the United States were trying many different ideas. The same Howard Gardner, who developed the theory of multiple intelligences, had long been involved in trying to establish whether there was any relationship between the arts and academic performance. A Harvard University professor, Gardner had been trained in music as well as psychology, and he felt that there was a connection between arts education and enhanced academic ability. From 1972 to 2002 he was codirector of Harvard Project Zero, an ongoing investigation begun in 1967 to study and improve education in the arts. The project was initiated by the philosopher Nelson Goodman, and it was named Zero because of his belief that at that time, nothing had been firmly established about the link between the arts and cognitive learning.

Not long ago, Harvard Project Zero reviewed 11,467 articles, books, theses, conference presentations, technical reports, unpublished papers, and unpublished data. One hundred eighty-eight reports investigating the relationship between one or more arts areas to one or more academic areas were combined in a meta-analysis. The results, described in the REAP (Reviewing Education and the Arts Project)[21] report, indicate causal relationships in only a few cases:

- Listening to music improves spatial-temporal reasoning

- Learning to play an instrument improves spatial-temporal reasoning

- Participating in play-acting improves verbal skills

- Learning to dance improves spatial reasoning

No other causal relationships were found, though such relationships were not definitively ruled out. Many correlations were suggested but not proved.

I have no doubt that there are connections between the arts in the curriculum and enhanced or even superior academic achievement. One possible explanation is increased motivation to learn academic subjects, as well as increased self-confidence, perseverance, and general well-being.

Using the arts as an entry point into academic subjects certainly captures young minds and stimulates creative thinking. This observation suggests a correlation between arts in the curriculum and academic success but does not prove any cause-and-effect relationship. I'm not familiar with each of the dozens of arts-in-education programs assessed in the REAP project, but my sense is that our musician residencies are making more direct connections between arts and academics than most school programs.

An especially important model for the quintet's curriculum at Bolton was the Every Child A Winner program, based on a theory developed by Rudolph Laban, a choreographer, dancer, and movement theorist who lived from 1879 to 1958. Every Child A Winner is a physical education program that stands in stark contrast to the competitive star system that so often prevails in school sports programs.

Robert Franz immediately identified with the philosophy of Every Child A Winner, which takes a very broad view of what education in movement can teach. In this program, children learn about how their bodies work and also about scientific and mathematical concepts, social relationships, and creativity. Along the way they gather "world knowledge" about architecture, history, and many other subjects. Every Child A Winner stimulates critical thinking and contributes to self-esteem. The program is carefully sequenced so that all the activities are appropriate to the developmental ages of the children.

Traditionally, training in music and training in sports have some similarities. To children who are not naturally gifted in these areas, physical education classes and music lessons are often dreaded occasions. The humiliation that many children suffer at having their clumsiness and ineptitude displayed so visibly can last a lifetime.

The Every Child A Winner method is a gentler approach. It grows out of the idea that movement is basic to all human activity, including intellectual achievement. Programs like Every Child A Winner aren't aimed at producing athletes—the idea is that education in movement will be helpful to future doctors, lawyers, artists, teachers, programmers, and pilots. This holistic movement-education program proved to be a sturdy bridge to our holistic music program.

Similarly, the quintet's curriculum presumes that music is a fundamental human activity. It too is grounded in the child's awareness of his or her own body. And likewise, it excels at developing general problem-solving ability in children.

Far from giving the children a little rest and recreation, the quintet immediately began providing a rich brew in which everything was connected to everything else, and every encounter was a lesson in listening, abstract reasoning, problem solving, new terminology, social skills, and more.

"With a good teacher," Robert said, "everything becomes a lesson."

CHAPTER 6

Models and Mentors

"I would really love to do that part slower," Cara says, as the quintet comes to the end of a passage it is rehearsing in front of the class.

"I think the way we're doing it is about right. I don't think this passage would work any slower," says Bob.

"Why don't we try it both ways?" suggests Lisa. "Let's listen to how it sounds and then decide."

A little later in the lesson, the quintet plays a piece with all five musicians playing at the same volume.

"Which instrument was playing the melody?" Lisa asks.

The sixth graders can't answer that question. What they heard didn't sound like music, and it was impossible to pick out the melody.

The quintet then plays the piece correctly, with the sound of the instruments balanced, and with different instruments picking up the melody in turn.

This is part of a lesson called Open Rehearsal, which we had first used with middle school students as one of several units on teamwork. It would be hard to think of a better model of teamwork than a chamber music ensemble. Like a professional basketball team, a classical music trio or quartet or quintet is a high-performance team, in which the members are interdependent and mutually accountable for the result. Likewise, they are a work group of which it can be said that the whole is greater than the sum of the parts.

EXPANDING TO A MIDDLE SCHOOL

Soon after we started the music residency at Bolton, the quintet began a three-year residency at Hill Middle School in Winston-Salem. Patricia Holiday was the principal at the time.

"Ann Shortt was one of my mentors," Pat Holiday says. In terms of its demographics, "the population at Bolton was almost identical to the student population we had at Hill. The state had already begun to focus on accountability testing, and we were looking for ways to help our children be successful."

Dr. Shortt told Mrs. Holiday about the results she was seeing from the quintet's presence at Bolton, and Hill was able to initiate a similar program with the same group of musicians, through a donation from its business partner, the Pepsi-Cola Bottling Company.

"My dream was to expose those kids to a world beyond what they lived in," says Mrs. Holiday, "and our business partner responded quickly and generously."

At Hill the quintet worked very closely with the classroom teachers to tie lessons to the broader and more advanced middle school curriculum, which included more science and social studies, for example, than the primary grade curriculum. One thing the sixth-grade teachers requested was a unit on teamwork and collaboration.

As Pat Holiday points out, people are interdependent and must cooperate and collaborate. "That's something we must teach children," she says.

The musicians are experts in giving each other's ideas a fair hearing, in negotiating differences, and in sometimes leading, sometimes following to get the best results.

Of the many different learning styles that have been identified, collaborative learning is getting more attention these days. In my school days, there was a bit of a stigma about students helping other students with homework or with finding the answers in academic assignments. Collaborative efforts were more likely to be condoned or encouraged for art projects and extracurricular activities. Also, there was some sense that when students worked together, particular students would always be the leaders and teachers of others.

Today, with the emphasis on diversity, group assignments are more often designed to point up the value of having a variety of skills and viewpoints to bring to a task. They can teach children that two—or three or five or six—heads are better than one, when what's in each head is different. Our quintet models that concept vividly.

It's also true that many schools today are making concerted efforts to combat violence, bringing a wide range of resources to bear on this thorny problem. Here again the quintet demonstrates how to disagree without resorting to hostility, bullying, name-calling, or blows. Chamber ensembles don't survive if their members can't negotiate their differences without damaging their relationships. They break up, as marriages and business partnerships do, if they don't learn to resolve conflicts in mutually acceptable ways.

Because of the emphasis in the wider world on collaboration, diversity, and peaceful resolution of conflict, I predict that our musicians will be asked more often to go into classrooms to "show and tell" the skills of teamwork. But even when the musicians are not demonstrating their own processes so explicitly, they are always acting as models. Because of their musical training, and because of the kinds of lives they lead as professional musicians, I think the role models they provide are different from what children usually see in the adults in their lives.

Here are some of the characteristics of professional musicians that our quintet members consciously or unconsciously bring into the classroom:

A sense of vocation. It's common for musicians to say that they didn't choose music as a career but that music chose them. More often than not, they go into this profession over the fears and objections of their families, and with the knowledge that wealth and financial security are not in store for them. More often than not, they know at an early age, much earlier than their peers in high school and college, what they are going to do with their life.

When we started the Bolton project, most of the quintet members were in their mid-20s, yet they had a sureness of purpose that is uncommon among young adults. All five of them had begun playing their chosen instrument by age 15. I myself started conducting at 13, and knew by age 15 what I wanted to do with the rest of my life.

A creative approach. Like all artists, musicians are creative and enjoy exercising their creativity. To the original members of the quintet, the challenge of coming up with a curriculum for three different grade levels, pretty much from scratch, was exhilarating.

Elizabeth "Lisa" Ransom, the quintet's flute player from 1995 to 2002, and director of the Symphony's education programs after Robert Franz moved to Louisville, was 24 years old when we started the project. Already she had considerable experience in inventing ways for musicians to be more involved in the community.

"At the North Carolina School of the Arts, Robert Franz and I were always wanting to do more," Lisa recalls. "I managed the West End Chamber Ensemble, which is one of the groups we started when we were students. I booked concerts for us and started to design educational programs. We felt strongly that outreach and education were an important part of our mission as musicians. We were young and energetic, and when the Bolton project came along, it was very stimulating. We all felt creative and free to try things. We had the feeling that there was nothing we couldn't do."

Lisa, Robert, and the other quintet members are representative of the young generation of classical musicians who expect to have to market their talents and make their own way in the world, looking to a career that consists of an ongoing series of projects they help create, rather than a steady job. Lisa credits saxophonist James Houlik, who taught an innovative career development course at the North Carolina School of the Arts, as her inspiration for this entrepreneurial view of the musician's life. She is a fine example of it. As of this writing, she teaches flute at two colleges, regularly plays in three orchestras, performs in small chamber ensembles—and teaches that career development course at the School of the Arts now that James Houlik has moved on to Duquesne University. During the summers, she teaches flute at a music camp in Maine. She has also held administrative staff positions with both the Winston-Salem Symphony and the Greensboro Symphony.

This creative approach to life translates to a high degree of flexibility in the classroom. Just as there are endless ways to put together a career, there are always more ways of helping children grasp concepts and learn skills. If one approach doesn't work, the quintet members thoroughly enjoy finding other ways, other analogies, other activities that help make an idea clear.

A different model of the teacher-student relationship. "I am always searching, always learning," Lisa says. "I think maybe this attitude came from Philip Dunigan, the most important influence of my life."

In talking about her famed flute teacher, Lisa could be almost any professional musician. In the short biographies used in performance playbills, musicians almost invariably name their teachers. In talking about themselves and telling their stories, musicians always tell you whom they studied with. No matter how old they are, they never lose their sense of connection to the teacher-mentors who helped develop their talent.

Music students will move across the world to study with a particular teacher, and teachers, in turn, have a strong sense of their influence on and responsibility to their students. The relationship between student and teacher is intense, one-on-one, and likely to continue for several years.

The individualized method of learning, by which teachers hand down to their students what their own teachers taught them, is part and parcel of musical training. Musicians can usually trace their musical antecedents through their teacher over generations to a great guru or mentor. Thus, I had most of my early music training from my mother, who had studied with the legendary Robert Teichmüller in Leipzig, who in turn taught her what he had learned from Clara Schumann, one of the leading pianists of the nineteenth century. Musical ideas and traditions are passed down this way, person to person, across continents and centuries.

Even though our musicians in the classroom do not spend enough time in the school to forge deep personal relationships with the students, I think they do carry a sense of the almost sacred role of the teacher.

A commitment to continuous improvement. "Being a musician brings the understanding that everything is a process," Lisa says. "We feel free to draft and revise, and to keep making it better as we go along. At some point it occurs to every musician that this is a process that will never end. There's always something to improve."

For musicians, knowing that further improvement is always possible is an exciting thought rather than a discouraging one. Musicians are always raising the bar for their own performance. No matter how accomplished they become and how much recognition and reward they receive, almost all professional musicians remain open to coaching from the masters and committed to making incremental improvements to performances that sound fine and flawless to most listeners.

Taken to school, this attitude translates into high standards and high expectations. The musicians take it for granted that the students—whatever point they start from—will be able to learn, and to keep building on what they have learned.

Knowledge of how to give and receive feedback. For the most part, musicians can take constructive criticism without damage to their sense of personal worth. Their whole musical instruction is a kind of feedback mechanism. Their music teachers constantly give them feedback on the smallest details of their playing, with suggestions on how to improve each aspect. In rehearsal and performance, they get regular feedback from the conductor, if playing in an orchestra, and from the other members of the ensemble, if playing in a chamber music group. Audience reaction is, of course, another form of feedback. Part of the risk taking of being a professional musician is the willingness to accept feedback, in front of one's colleagues.

Most professional musicians also have music students of their own, and they train them in the best ways that they themselves were trained, encouraging them and trying to help them become more skillful. It is part of our credo that we will transmit our knowledge and experience to our students.

A musical education seems to promote a different notion of how self-esteem is instilled than the ideas that are currently advocated in child-rearing and educational circles. Most musicians, if they thought about it, would probably say that self-esteem and self-confidence stem from being able to do something well. The better prepared and rehearsed they are, the higher their self-confidence.

Our musicians in residence carry this belief into the classroom. They don't think of children's self-esteem as so fragile that it will be shattered by the suggestion that the child guessed wrong or jumped to an invalid conclusion. They make corrections matter-of-factly, with no feeling that a child is a failure because she has made an error, but with confidence that the feedback will help the child learn and be accurate the next time. From their own experience, they assume that being guided in how to get it right will do more for a child's self-esteem than meaningless blanket approval of everything the child says and does.

A focus on the realities of the present moment. To play live music is to live and work in real time. If something or someone is missing, if a mistake is made, if one player comes in too soon or too late—the show must go on! You can't unring the bell. By the same token, we musicians are quick to take advantage of any favorable circumstances or unexpected but welcome events. Performing makes realists out of performers. We don't spend much time bemoaning the fact that conditions are less than ideal, and we can't get away with blaming others and making excuses when our performance is not as good as we would like.

The ability to deal with what is in front of them—not with what they planned for, not with what they would necessarily prefer—is something the musicians in residence constantly exhibit and model. The last couple of sessions at Bolton Elementary, in the spring of 2002, are a good example. The classes were scheduled for May, when the whole school was in an uproar because of the dreaded end-of-grade tests, on which the school and the teachers would be rated, ranked, and rewarded. When the musicians arrived at the first classroom, it was empty. Under the pressure of test-taking preparations, the teacher had simply forgotten about the quintet's visit. As the class was notified and the children were on their way back, the musicians quickly adjusted their lesson plan to fit 20 minutes instead of the expected 30 minutes. At the next visit, many children were receiving intensive one-on-one tutoring, and the musicians were told that their presence would be an interruption. They took the rest of the children outdoors on the lawn, set up chairs for themselves, and modified their teaching to cope with passing traffic and other environmental disruptions.

Playing music teaches people to make little course corrections all along, and to face larger disruptions with ingenuity and equanimity.

A passion for excellence. The musicians don't spend the major part of their time in the classroom actually playing music, but when they do play, the sound is exquisite. Part of what keeps the children's attention focused is that the musicians play so well. The sound is crisp, clean, balanced, fluent, and expressive all at once, and its quality simply commands attention.

You don't have to be musically sophisticated to respond to music of any kind played expertly, just as you don't have to know or care about sports to respond with pleasure to the feats of Olympic athletes. This kind of mastery is immediately recognizable. This kind of excellence impresses and inspires. It acts like a tonic on everyone in the vicinity.

The musicians also talk about their commitment to always playing their best. In every class, a child will ask how they play so well, and the musicians talk about how much they practice. They describe how they keep working on the hard passages so that they won't be afraid when they have to play them in front of people.

The musicians are an inspiring model of high achievement, and a great source of information about the habits and attitudes that contribute to excellence.

The sense of proportion that comes from serving something larger than themselves. When you ask musicians why they do what they do in spite of the sacrifices they usually make, words fail them. Unlike most jobs and careers, music involves every aspect of one's being—intellectual, physical, emotional, and spiritual. No matter how much they learn and how dazzling they become in performance, musicians never come to the limits of music. There is always more to learn and more to express.

Robert Franz says that "seeing adults who love what they do" is one of the main factors in the success of the music residencies. I think he is describing not just adults who enjoy what they do, or have fun making music. I think he is talking about the lifelong passionate attachment of musicians to music. And I agree that the children do get a sense of that when musicians are regular visitors to their classrooms, and that it is inspiring.

Although each of these traits and skills can be learned through other disciplines and experiences, the schools, in getting five carefully chosen musicians who essentially function as a high-performance team, receive a concentrated dose of the character that musical training builds.

Some of it does rub off, on the other adults in the school as well as on the children. I believe that when the quintet works with many of the same children and the same faculty members for several years, the musicians become significant role models and mentors. And what they are modeling is by no means irrelevant to high academic achievement—or to the ability to navigate and succeed in the world of work.

"Do what I say, not what I do" is many a parent's rueful acknowledgment that modeling a behavior would be the best way of teaching and instilling it. Musicians in the classroom may be proving in their own way that there's no business like "show" business. These musicians teach by example!

CHAPTER 7

Learning to Listen

"Imagine a landscape," Bob Campbell says. "What are some things you might see in the landscape?"

The second graders' answers come quickly: sky…sand… trees…ocean…grass.

"The ceiling? That's more of an indoor thing, isn't it?" Bob comments on one child's response. "What would be something you would see outdoors, in nature?"

When it's clear that the children have grasped the notion of a landscape, Bob says, "When I think of a landscape, I think of that great big painting at Reynolda House.[22] Have you seen that? Paintings of outdoor scenes like that are called landscape paintings. Now we're going to paint a *soundscape*.

"Let's say it's a summer day, and you're walking out in the country, beside a creek. What are the things you would hear? I want you to use your imagination."

Coming up with a variety of sounds is a little more difficult, as the children are less accustomed to using their sense of hearing to describe an environment.

"Water?" a child ventures.

"Yes, the sound of water moving."

"Wind?"

"Yes, the sound of wind blowing."

Eventually the list grows to include the sounds of leaves rustling, frogs croaking, crickets chirping, bees buzzing, and snakes rattling. Then Bob asks the children which of the instruments in the classroom could best make each of those sounds.

With a little trial and error, the children hear how the flute can sing like a bird, the bassoon can mimic a bullfrog, the clarinet can be persuaded to buzz like a bee, and the gentle clacking of the oboe's keys might sound like rustling leaves.

One of the quintet members writes these correspondences between instruments and elements in nature on the board, to give the children another way to help fix the relative sounds of the wind instruments in their minds.

Then Bob asks the children to close their eyes and imagine themselves taking a walk in the country, beside the creek. Bob draws their attention to a calling bird; a chirping cricket; a croaking frog; and the sounds of wind, water, and bees. Debi, Eileen, Cara, and Kendall produce these sounds on their instruments. Now the children are using their ears only, not their eyes, to commit to memory the sounds of the individual instruments.

"Uh-oh, I think I hear a rattlesnake," Bob says. "Let's get out of here."

He stamps his feet and the children follow suit.

This exercise in imagining a soundscape is the second of three soundscape activities that the quintet normally presents in one half-hour session. Together these three exercises seem to exert a powerful effect on how children listen.

In the first exercise the kids sit with their eyes closed for a timed minute, instructed to notice any noises and sounds that occur. When the minute is over, they tell the quintet what they heard. The list is typically extensive: the sound of the heating or cooling system, footsteps and talking in the hall, a car going by the building, paper and clothing rustling, a door closing, a breeze blowing, floorboards creaking, shoes squeaking, someone sneezing, a neighbor yawning, a stomach growling, and the sounds of their own breathing.

Members of the quintet say that some children today have little or no experience of this kind of quiet. In their home environments, television sets, radios, and other sound systems are always on. For some kids, this quiet minute represents the first time they have consciously been aware of small background noises.

In the third exercise, the children create a soundscape. First they think of sounds they might hear during, for instance, a thunderstorm. They may come up with wind, thunder, rain splattering on housetops and roads, the distinctive sound of car tires on wet pavement and leaves, the sirens of emergency vehicles, the howling of dogs. Lightning is invariably named, and the children need to think it through to realize that lightning has no sound of its own, though an object being struck by lightning may give off a sound.

The musicians then divide the class into five groups, with a member of the quintet to guide each group through the activity. One group might be wind, another thunder, another raindrops, another sirens, another dogs. Eileen then tells a story, of a day that started out pleasant and then turned stormy. As thunder and wind, sirens and barking dogs come into the narration, the appropriate groups make their assigned sounds, becoming louder or softer according to what Eileen is describing.

The quintet discovered that 30 minutes is often a long time for young children to sit still. By introducing some physical activity into each half-hour session, they are able to keep the youngsters listening and learning. The soundscapes session, which is done with increasing complexity and nuance at each grade level, has turned out to be pivotal. It changes the way children attend to sound.

We take hearing for granted. Our ears are never really turned off, even when we are sleeping. There is even evidence that people hear and can recall what was said in an operating room while they were under anesthesia.[23] Although we are taking in information through our ears all the time, most of it is "in one ear and out the other." We hear, but we don't pay much attention. We hear, but we don't listen.

A couple of weeks after we began our musician residency project at Bolton Elementary School, the classroom teachers began commenting on improved listening and longer attention spans. Almost three years later, when we got the end-of-grade test results—showing that the children had scored some 50 percentage points higher than had the children the previous year, who'd had no exposure to the musicians—I wondered if improved listening skills were somehow contributing to the improvement in test scores.

The following year, we applied for and received a grant to have education researchers from the University of North Carolina at Chapel Hill assess the residency program. This group of researchers had experience assessing the A+ Schools, a network of public schools across the state of North Carolina in which the arts are deeply integrated into the curriculum. Doctoral students under the direction of Dr. George Noblit, director of Evaluation, Assessment and Policy Connections (EvAP) at the University of North Carolina's School of Education, came to Winston-Salem to observe the Bolton project. In their report, the researchers concluded only that further study was needed. They observed that the musicians did many of the things that good teachers do. They speculated that the success of the project might be attributable to the "Hawthorne effect." This term refers to the finding that intervention of any kind may produce some improvement in a situation. In other words, as with the placebo effect in medicine, it is the special attention itself that produces the result, not the particular changes that are introduced.

I felt fairly sure that there was more going on than the Hawthorne effect. The school itself was changing under the leadership of Ann Shortt, who had introduced a number of innovative and effective programs. But I had taken those innovations into account. Children who hadn't received the music-enriched curriculum improved in academic performance by no more than about 20 percentage points. Only the third graders, who had had visits from the musicians over a period of three years, made the major leap on standardized tests. I decided to embark, on my own, on the "further study" that was needed.

I contacted the department of Neurology at Wake Forest University's medical school, then known as the Bowman Gray School of Medicine. When I described the information I was looking for, I was told that Frank Wood was the man I needed to talk to. I didn't know him at all.

Dr. Frank Wood, head of the Section of Neuropsychology at Wake Forest University School of Medicine, specializes in the neuropsychology of learning, including dyslexia. When I reached him by phone he was immediately interested in my questions and suggested that we meet for lunch that day.

Over lunch, as I responded to his questions about what the musicians actually did in the classroom, a growing smile of recognition played across Dr. Wood's face. As I was talking about what was going on in the classroom, he was thinking about what was going on in the brains of the young learners. The rise in academic performance made sense to him, given what he knew about how children learn to read, and what kinds of things may impede their learning.

He told me at that first meeting that the musicians were evidently improving the way the children listened. This sounded exactly right to me because we were calling their attention to very subtle distinctions in pitch, tone, duration, and loudness. Dr. Wood and the medical school have been involved in monitoring, measuring, and supporting our musicians-in-residence programs in the years since then, but that first meeting was the Eureka! moment for me. As soon as he mentioned the importance of listening, I sensed the truth of it. I felt certain that music was helping the children learn to listen in a particular way, and that better listening probably set the stage for them to be better learners.

Tuning in to the "sounds of silence" is a way to move from simply hearing to actively listening. In the musician residencies, our primary grade students discover that silence is full of sound. Small sounds. Tiny distinctions. Impressions that register only when you are deliberately listening for them.

Learning to read depends, among other things, on being able to hear and discriminate among little sounds. It depends on being able to distinguish phonemes, which are the smallest units of speech. For example, in the common word *cat*, there are three phonemes: /k/, /a/, and /t/. Every human language has its own set of phonemes. Think of the throaty sounds of German, or the soft, swishing sounds of Portuguese, or the demands that the "r" sound in Spanish makes on the tongue, for instance. American English has approximately 44 phonemes; some African languages have as many as 250.

Babies are born with the ability to distinguish all the sounds of every language in the world, the whole huge panoply of phonemes. Before learning to talk, however, the infant, by the age of 12 months, is becoming more specialized in regard to spoken language.[24] By age one the ear is already being tuned to the specific phonemes that make up the primary language, or in some cases two languages, that the child hears regularly and will learn to speak. An American child, for example, will not be able to distinguish among all the clicks, clacks, and whistles that are among the phonemes of some African languages. Those sounds will be "foreign" to the child, and hence difficult to reproduce or remember. This is particularly true as the child matures, and becomes even more difficult in adulthood. Notice how easily children pick up dialects and accents and how hard it is for their parents.

The essence of learning to read is matching one sense, hearing, with another sense, seeing. The first-grade child must understand that the letters on the page correspond to the sounds of speech. Children who are unable to match the sight and sound of words are called dyslexic. In the most severe cases dyslexia is usually a genetic condition that prevents children from distinguishing and manipulating phonemes.

Seeing, hearing, and processing are all required for reading, but there is no single "reading center" in the brain. Reading is like having to go to the butcher shop, the farmers' market, the bakery, and the wine shop to get the ingredients for tonight's dinner. The things you need to cook the meal are located in different parts of town.

From listening to the sounds of silence, children in our residency program move on to listening to their imagination and searching their memories for meaningful sound. In a later lesson, the musicians play a piece of music and ask the children what the music reminds them of. Perhaps they hear a kitten playing, or soldiers marching, or a giant laughing. Using the images and feelings suggested by the music, the children make up a story, draw a picture, or translate the music into gestures. Still later, as their listening skills improve, the children make pencil marks on paper to indicate the pitch, duration, and intensity of notes the musicians play for them. Or they may use colors to portray the emotions evoked by the sounds. Multisensory, hands-on learning experiences create more associations to particular sounds, and more connections within the brain.

All these activities reinforce the relationship between sound and the other senses. They virtually force the children to use their ears in ways that have not been emphasized in their earlier learning. The effect is quite rapid and dramatic: within just a few weeks, the classroom teachers notice that the children are actively listening. Their attention spans improve, and they seem better able to take in and retain information, and to follow instructions.

BOTH SIDES NOW

In 1861 a French physician named Paul Broca observed that damage to the left hemisphere of the brain was associated with loss of speech. Damage to the right hemisphere produced no such obvious and drastic loss. It is only in recent years that the right hemisphere has come to be appreciated for its different but important functions.

I recall one of my father's best friends, Arthur Benton, today a well-known neuropsychologist, who with his musician wife often came to dinner at our house when I was young. During these visits, we had lively discussions about the burgeoning field of hemispheric lateralization—that is, the study of the right and left hemispheres of the brain. He told us about studies of epileptic patients whose only hope to avoid massive seizures lay in the severing of the corpus callosum, whereby the two hemispheres of the brain were detached. Not only did the surgery stop the seizures, but patients led surprisingly normal lives, to a point: many reported differences in how they perceived the world about them. Clearly, the surgery had also severed some of the interdependence between the right and left hemispheres.

That interdependence is particularly striking in the way we process music. In *A User's Guide to the Brain,* John J. Ratey, M.D., professor of psychiatry at Harvard Medical School, points out that the left side of the brain is superior at attending to rhythm, while the right hemisphere is activated by the timbre, or quality of the sound. These were also the findings of PET scan studies conducted by Henri Platel at the University of Caen, France, in which men without musical training listened both to familiar pieces of classical music and to random sequences of notes.[25] The well-known music activated the part of the left brain that is usually associated with the processing of speech. Rhythm changes in the random sequences lit up the same area of the brain. Platel concluded that appreciation of music is enormously complex, involving the coordination of many specialized regions of the brain.

New evidence is emerging from studies of people who were born missing the ability to process and appreciate music, and from people who have lost that ability, through injury—a condition called amusia. Recent studies of people with amusia suggest that the regions for processing speech and the regions for processing music are not identical. People whose musical abilities are absent or destroyed often have no impairment to their language functions, and vice versa.[26]

These early findings are intriguing. The implication is that a portion of the human brain may be reserved for musical use. If that is so, the further implication may be that music may have had evolutionary value; otherwise, distinct neural pathways for processing music might not exist.

Functional magnetic resonance imaging is also showing us that not everybody listens to music the same way. The brains of trained musicians process music differently than the brains of unsophisticated listeners. In the early years of my marriage, I had plenty of firsthand experience of the different ways trained musicians and others listen to music. My wife and I enjoy entertaining, and we often give small dinner parties. Debra loves music and sings in a barbershop quartet, but she has never had formal musical training. She is expert at attending to the details that create an inviting atmosphere, and for her that includes pleasing music. Since many of our dinner guests are classical musicians, she used to like to have classical music playing softly in the background.

The problem is that for musicians, there's no such thing as "background" music. Debra's attempts at setting the mood with music could be counted on to stop conversation cold.

Every musician in the room would stop talking and listen attentively and actively to whatever was on the CD player.

Similarly, you can usually spot the trained musicians at noisy parties. We're the ones who are pretty much useless at cocktail chat. We're trained to listen to the whole gestalt, and it can be difficult or impossible for us to zero in on a single conversation amid the general din.

Almost all children in the primary grades fall into the musically unsophisticated category. They enjoy music as a total experience. In every lesson, our quintet asks children to pay attention to particular elements and aspects and qualities of music: Which instrument played the melody first? Which played the melody second? Which instrument should be the sound of the wind? With your eyes closed, can you tell whether the oboe or the clarinet is playing? Was that note higher or lower? Do you expect the next note to be higher or lower? Did Debi play fast or slow? Did Bob play that loud or soft? Why did that music make you think of a parade?

The children are no longer just passively listening; they're listening *for* something. They're using their ears to help them anticipate things, remember things, compare things, and imagine things. They are drawing on the complexity of classical music to fine-tune their ability to hear. They are drawing on more of their brains to listen and to learn.

CHAPTER 8

A Symphony of Neurons

"Playing in an orchestra makes you feel that anything is possible," says Bob Campbell, the quintet's horn player from the beginning of the Bolton project.

Bob was speculating about why the quintet has been able to reach and teach children who seem difficult or "impossible" to other adults. In the classroom, Bob radiates this sense of possibility and confidence to the children. Not only is he an extraordinary horn player, he is as handsome as a movie star. His relaxed manner and cheerful outlook are as appealing to the children as the sight and sound of his brilliant, beautiful horn.

Bob is old enough to have college-age children himself, and he has played in professional orchestras for more than 20 years. Yet his awe at the scope and majesty of a symphony orchestra is undiminished. I think most orchestral players share this feeling, and it's a good part of what keeps them going, despite long and irregular hours, grueling travel, and small paychecks.

The modern symphony orchestra *is* awesome. It is the most elaborate outgrowth of the human ability to express oneself in sound, the culmination of human effort to bend sound into beauty. You don't have to be a symphony conductor to regard the orchestra as a pinnacle of human achievement. First-time concertgoers marvel at the experience, and little children can be depended on to thrill to the sound of the orchestra.

An orchestra is made up of five families, or sections: the strings, the woodwinds, the brass, the percussion, and a miscellaneous keyboard section, which may include a piano, organ, harpsichord, or celeste. All the families make their sounds in different ways, and within each family are instruments of various shapes, sizes, and sound characteristics. The range of what this group of 100 or so instruments can produce is astounding. From Tchaikovsky's rousing *1812 Overture* to the grandeur of a Beethoven symphony to the haunting sound of Gershwin's *Rhapsody in Blue* to the pop rhythms of the latest Broadway show, the orchestra is the source of endless aural pleasure.

The secret to this infinite variety is in connections and relationships. The instruments can combine in myriad ways to achieve different effects and accomplish particular purposes. The orchestra is a network, in which the parts are interrelated and can be pulled together in any number of configurations. The rich, lush sounds of only the strings playing together, the bold and bawdy brilliance of a brass fanfare, the haunting and sinuous insinuations of an oboe or a flute, and the punctuation of the percussion are just some examples.

The orchestra is a relatively modern creation. Before the 1600s, musical groups banded together for dances or to accompany troubadours and minstrels. Starting in the seventeenth century, the strings became the backbone of the orchestra, and the gradual addition of wind instruments supplied strength and variety. In the eighteenth century, Haydn and Mozart codified the size and composition of the orchestra, as well as the musical form now known as the symphony. In the nineteenth century, with the rise of Romanticism and the desire to let it all show, color, or richness of sound, became as important as structure. The medium became the message in a sense, so the orchestra got bigger and more varied. In the twentieth century, electronic instruments and amplification, echo chambers, and the recorded elements became orchestra players. Now orchestral music has become ubiquitous, often furnishing the sound of television commercials, shopping centers, and video games.

The orchestra resembles another network, which we all are born with—our human brain. The more we learn about the brain, the more we understand that its whole capacity is a matter of making connections between neurons, or brain cells. Joaquin Fuster, a professor of psychiatry and neurobiology at the University of California at Los Angeles and one of the foremost theorists of brain science, believes that all knowledge is encoded by relations, and thus by connections, in the neural networks of the cerebral cortex.[27] The way our brains acquire and store knowledge is by adjusting the transmitting capability of the connections, or synapses, between neurons.

There are times, or windows of opportunity, during which connections for learning are best made. The ability for learning particular skills at particular ages has a physical basis. The human brain comes into the world only partly developed but ready to be "hooked up." It needs interaction with the external environment—sensory input—to complete its development.

This characteristic of our brains is something that sets us apart from most of the animal kingdom. Our closest relative, the chimpanzee, is born with a brain the same size as a human baby's brain, about ten ounces in weight. Whereas the chimp's brain will not grow significantly after birth, the child's brain will continue to grow for about a dozen years, quadrupling in size. The chimp is able to walk at about one month old and is fully mature by nine months. Maturity comes much more slowly to human beings and requires input from the world.

Everyone who has ever had a child, and many people who are not parents, have a sense that there are times when children are ready and able to learn certain things. Toilet training, telling time, and tying shoes are common examples. If the child is ready, these skills are quickly learned. If not, the process is long, frustrating, and frequently unsuccessful.

Similarly, most people have a sense that there comes a time when it is too late to learn certain kinds of things, or at least to learn them as well and as easily. People who move to foreign countries, for example, struggle to learn the native language but see their young children pick it up as though by osmosis. Teenagers and adults, no matter how great their efforts, rarely learn to speak the new language with a flawless accent, as a young child will do without effort.

This aspect of learning is important in the context of educating children because we can't assume that they will automatically acquire various skills and knowledge. We human beings are "hardwired" for some skills, just as animals are. Babies will suck, cry, hold up their heads, sit, crawl, and learn to walk with or without encouragement and instruction. They will learn to speak if they are around human beings who speak. But human beings are not genetically programmed to read and write. These and many other skills require considerable interaction with the external environment; they also require the child to have reached a certain level of development. A one-month-old baby cannot be toilet trained; a one-year-old baby is insufficiently developed to learn to tie her shoes or read a book.

Our brains remain flexible throughout life, continually being restructured by new things we learn. In *A User's Guide to the Brain*, John J. Ratey notes that the brains of children three to ten years old use twice as much glucose, or fuel, as adult brains.[28] That is because vast numbers of connections are being made during those years.

In the young brain, weak connections—those that have not been reinforced by repetition—get pruned out. Other connections get bigger and stronger from repeated use. Ratey writes that magnetic resonance imaging shows that the brains of violin players have more area devoted to the pathways representing the fingers of the left hand. These are the digits that get the most exercise during the musicians' hours of practicing. Every human brain is different from every other, and different from one year to the next, depending on what we are "feeding" our brain by what we learn and what actions we perform. Experience is always restructuring our brains.

The woodwind quintet is similar to a group of interconnected brain cells within the larger network. We reinforced the connection of these five people to one another by sending the quintet to the school, where the members spent many hours practicing, playing, making decisions, and teaching together. At first they were not terribly efficient, because the pieces of music they played together and the team teaching they were called on to do were both new to them. But the more times they played Mozart's *German Dance*, for example, the easier it became. The more they spread the teaching of lessons among the five of them, the more automatic this way of working became.

Similarly, residencies of several weeks in a school gave the children the repetition that is needed to make strong connections—that is, to learn. We can imagine, for example, that hearing the sound of an oboe once might make little or no impression on a young brain. But hearing an oboe on 12 or 15 occasions, closely related in time, might well be enough to stimulate some new connections. And if the oboe is being presented as a woodwind instrument, as a double-reed instrument, as an instrument that can sound like a cricket chirping, as the instrument played by Cara, and as an object made of wood and metal, then many new connections are being created and strengthened.

In the orchestra, and in the quintet, Cara and her oboe will be doing different things at any given moment, depending on the piece of music in front of her. They may be reinforcing the theme of the piece. They may be accompanying the main melody. They may be performing alone or with a few other instruments or with the entire ensemble. They may be resting and waiting to come in. Within a single concert, or even a single classroom lesson, the oboe will take its place in creating many different sounds, evoking many moods and meanings. Everything the oboe does is tempered by all of what's going on in any given moment.

That is the way our brains work too. Only a few things have specific, set locations. The controls for our body movements are located in the motor cortex, a band that runs from side to side across the top of the head. Vision, touch, and the other senses also have their own locations, which vary little from one person to another. But the *whole brain* is involved in most activities, including perception, attention, planning, and memory. A piece of information like "oboe" doesn't just exist in one spot in our brains. It is "filed" under many different categories; it is part of many different neural networks and layers of neural networks.[29] It can be evoked by hearing the particular sound of an oboe, but perhaps also by hearing the word *woodwind* or *Mozart* or *Cara* or *quintet* or *Peter and the Wolf* or any of a variety of associations. Its particular meaning will be colored by what it is linked to within any of numerous neural networks.

The frontal cortex is one of the last parts of the brain to reach maturity. This is the part of the brain that helps organize and categorize information, discriminates one thing from another, and plans and sequences thoughts and actions. It is not fully developed or myelinated until after puberty.[30] Thus, the elementary and middle school children our musicians work with are at the developmental stage at which the rich and complex lessons that music can teach fall on fertile ground. The brains of these children are ready to grow by making more and more connections, and to become faster and more efficient at storing and sorting information. Learning to read is a clear example of the relationship between readiness and learning. Almost everywhere, children are required to attend school by age six or seven, because most children are able to learn the complex skills of reading at that age.

If children for some reason fail to learn certain skills while the "window" is open, the prospect of their ever learning them is bleak. Children who are raised in conditions of such isolation, abuse, and neglect that they never hear the human voice may never be able to speak well. This theme was explored in the movie *Nell,* in which Jodie Foster played an isolated young woman who'd developed her own form of language.

A similar phenomenon occurs when the physical obstacles to sight are removed in a person who was born blind. There is a window of opportunity for learning to see, and after the window closes no surgical procedure can confer the gift of sight. A child born with cataracts can learn to see if they are removed early. But that child will remain blind if the cataracts are surgically removed after the age of two or three.[31] The ability to perceive light and dark and motion may be attained, but the ability to make sense of what is seen will not, despite the perfect functioning of the eye itself. Not long ago, the media widely reported the story of a middle-aged man who had been blinded at age three and who had recently recovered the use of one eye.[32] He found it a hindrance rather than a help, as he had never formed images of the people or objects in his adult life. He cannot recognize his own wife until he hears her voice. A cube looks like a square with extra lines, and he must close his eyes to ski, a skill he had mastered while blind.

For most skills, however, the windows do not close with such a bang. There are stages in development when it first becomes possible to learn something, when the window of opportunity opens—and in many cases the window remains open. As many senior citizens can attest, the old dog *can* learn new tricks. Learning to swim, sing, bake a soufflé, or surf the Internet may come more easily to young people, but every day people in their 70s and 80s are trying their hand at these activities for the first time.

After a young child has learned to walk and talk, a major brain pruning occurs during which the connections necessary for many baby behaviors are discarded. The child will no longer need to cry to signal that she is hungry, or to throw a tantrum, or to crawl to reach a toy. The baby stuff goes, in much the same way that we relegate the crib, the high chair, and the stroller to the attic or garage sale.

Another big clearing out of brain cells occurs at puberty—generally age 11 or 12 for girls, and 12 or 13 for boys. It's likely that certain knowledge that has been constructed in our brains needs to be reconfigured in order for us to take on our role in ensuring the survival of our species. For the young child, the family is the place to look for love and nurturing. The young adult needs to be able to move out of the nest and look to a larger world to find a mate and become a parent. Puberty is when the window closes for acquiring speech for the first time, for learning a foreign language with the same fluency as the first language, and for becoming a musical virtuoso.

The human brain is by far the most complex thing on earth. We can hardly begin to fathom its intricacy. But orchestra musicians may have a visceral sense of how such nimble networks function because of their orchestral experience. At the very least, playing in a symphony breeds into them the knowledge that there are always many different ways to tackle a challenge, and that there is more potential within the orchestra than is evoked by even the full repertoire of symphonic music. Their different experience of limits and limitations may translate into how they view and teach children.

In brains as in orchestras, it seems that anything is possible.

CHAPTER 9

Can You Say "Legato"?

"We're going to talk a bit about opposites today," Bob says. "It's kind of hard to say what opposites are without giving some examples. Who can give me some examples of opposites?"

Most of the third graders raise their hands. And they can think of lots of examples. Left and right...Hot and cold...Up and down...Stop and go...Right and wrong...Wet and dry...Bald and hairy.

"We have some opposites in the quintet," Bob tells them. "Our opposites don't have so much to do with the way you see something, but more about the way you hear something."

Bob introduces four pairs of opposites found in music: high and low, fast and slow, loud and soft, bumpy and smooth.

Cara then takes over the lesson. "HELLO," she booms. "Was that loud or soft?"

"Loud!" the kids call out.

"We have fancy Italian names for loud and soft," Cara says. "When a sound is loud, we call it *forte*. When a sound is soft, we call it *piano*. Can you say *forte* in a forte way?"

"FOR-TAY!" comes the response.

"Can you say *piano* in a piano way?" Cara speaks in a whisper and holds a finger to her lips.

"Pi-ah-no," the children whisper.

"We have fancy names for bumpy and smooth, too. When sounds are bumpy, we say they are stac-ca-to. When sounds are smooth, we call them legato. Let's say *staccato* in a staccato way."

"Stah! Kah! Toe!"

"Now let's do it with our hands," says Cara. The children mimic her short, choppy hand movements.

"When the music is staccato, there's space between the notes," Cara says.

"Now let's say *legato* in a legato way. Use your hands, too."

The children smooth the air with their hands as they intone "lehgahtoh."

Eileen writes the names of the pairs of opposites on the board and asks for a volunteer to come and choose one from each set. Then Kendall, ostensibly following the child's instructions, plays a short passage on the bassoon.

"Was it forte?" Eileen asks.

"YES," comes the fortissimo response.

"Was it slow?"

The children agree that it was.

"Was it low?"

"Yes."

"Was it legato?" Eileen asks.

Yes, the children heard the smoothness of the sound. Another child picks four words from the pairs of opposites, and Debi plays them on the flute.

"Was it fast?" "Yes." "Was it legato?" "Yes." "Was it piano?" "Yes." "Was it low?" "No." In fact, Debi was trying to trick them by playing high notes.

Before the lesson is over, the children learn that not all the instruments can play all the opposites equally easily or well. The bassoon handles slow and loud better than the flute. The flute excels at fast and legato. And the nimble clarinet can go quickly from one extreme of its register to the other, from highest to lowest, and vice versa.

When it's time for the quintet to play its closing piece, Eileen instructs the children to listen for the opposites, when the music is high and low, fast and slow, forte and piano, staccato and legato.

The idea of opposites will be prominent in the education of these children all the way through school and into college. At every stage they will be asked to "compare and contrast," to say what is the same and what is different. Locating things on a continuum is a useful way to think about and describe our experiences. *Today is warmer than yesterday. That was the worst hot dog I ever ate. The movie is good, but it won't change your life.*

The swimmer's performance was a 7. Knowing the outer limits helps us define things quantitatively. Knowing the nature of a particular continuum enables us to define things qualitatively and to avoid "comparing apples and oranges," or trying to explain "what that's got to do with the price of eggs." The concept of opposites also helps us perceive change over time, as we compare the present with the past along a given spectrum. *I can't hear as well as I did when I was 20. These new computers are ten times faster than the ones that were produced a few years ago. We eat many more ethnic foods today than we did when I was a kid.*

Actually, at every grade level, the quintet members have introduced the idea of opposites from the very first lesson, when they spoke about their instruments in terms of high and low notes. And they are about to introduce another set of opposites. Just as there are high notes and low notes, there are high and low body parts, which the children are much more familiar with. Asking the children to mimic the music physically reinforces the kinesthetic learning mode (sometimes referred to as "muscle memory"). Lessons in opposites bring the kids out of their chairs—they raise and move their arms above their heads in response to the high notes of the flute, and stomp and shuffle their feet to correspond to the low notes of the bassoon. The quintet uses visual and kinesthetic exercises to help teach the auditory lessons of musical opposites.

A new concept, of a middle between the opposites, is taught with an instrument that is not as high as the flute nor as low as the bassoon. The children look like hula dancers as they move their middles to the tune of the clarinet; moving their arms and legs and other body parts as well helps the children get a literal *feel* for the musical opposites: fast and slow, smooth and bumpy, and loud and soft, in addition to the middle ground. And these middle grounds seem natural: medium soft movements are larger than piano, but not as large as forte.

Bob Campbell mentions to the third graders that it's difficult to define opposites without giving examples. What is really meant by opposites is the far ends of the same continuum. Bob helps the kids with this fairly abstract concept by asking what the opposite of cat is. The children always name dog as the opposite of cat. Bob then points out that cats and dogs may be opposite in some ways but that they have many things in common—whiskers, paws, noses, fur, and so forth.

GOING TO EXTREMES TO IMPROVE LISTENING

Cat and dog are not really opposites, of course. But even true opposites contain both similarities and differences. Hot and cold are similar in that they both describe temperature; up and down are similar in that they both refer to physical space. Music seems to be ideal for instilling this understanding, and for giving children a different way of grasping the meaning of opposites. Music is infused with the idea of opposites and the continuums they inhabit, from one end to the other.

Musicians speak of playing in the low register or the upper register of their instruments, where *register* is the pitch continuum, from the lowest to the highest notes the instrument can play.

When musicians use the word *dynamics*, they are referring to the degree of loudness of a passage, from *piano* (notated as *p*), through the *mezzo forte (mf)*, or middle range, to *forte* (*f*). Some composers may even denote the extremes of this continuum as *ppp* and *fff*.

Staccato and *legato* are at either end of the spectrum called *articulation*, referring to the spacing of the notes as they are played. The middle range is what string players would call detaché, where the individual notes are distinct but not sharply "bumpy." Wind players often use "dah-dah-dah" to describe the sound of the middle range, as opposed to the sharper "Tah Tah Tah" they use to describe staccato passages. Brass players may speak of the "pear-shaped tones" of this midrange, which conjures up the visual image of rounded, bottom-heavy fruits hanging individually from a branch. For singers, words generally dictate the articulation.

The speed at which all these things happen is the music's *tempo*. This continuum yields its own crop of fancy Italian names: *allegro* for fast, *presto* for really fast, *largo* for slow, and *adagio* for even slower. In the middle are *andante* and *moderato*. Tempo is one of the most important considerations when the conductor prepares a piece for performance. There will often be a particular passage in the music whose intricacies dictate a certain tempo. Certain staccato passages bounce well "off the string" at the right tempo; other doubled notes, dubbed "scrubbed" notes by string players, limit the speed at which they can be played. And always the "local" tempo (the speed of a particular passage) is subservient to and dependent on the tempo and pacing of the larger shape of the piece, whether a movement or a magnum opus.

To summarize this new vocabulary, *register* refers to high or low pitch (frequency in physics), *dynamics* to loud or soft (amplitude in physics), *articulation* to smooth or bumpy (duration in physics), and *tempo* to speed, whether fast, slow, or somewhere in between (velocity in physics).

In contrast to how they perceive "wet and dry" or "light and dark" or "stop and go," the children need to use their ears to understand musical opposites. A great deal of what the quintet does in the classroom helps refine the sense of hearing. Our bodily senses are basic to all learning and memory. Our brains use sensory input as the first element in forming what Joaquin Fuster has named "perceptual cognits." Cognits, he postulates, are representations in the brain of any item of knowledge—about the self, the world, or the relationship between them. They are the primary element of all cognitive activity and could be described as networks of associations. The seed of each one of them is something that comes to us through sight, hearing, touch (including such things as pain, heat, and pressure), taste, or smell.

I think parents typically have an intuitive sense that the way to learning is through the senses. If you listen to parents talking to babies and young children, their talk is most often about the look and feel and taste and smell and sound of things. Toddlers teach their parents that they do not know or care about abstract concepts or moral principles or even emotions. What's real to them is what they perceive through their senses.

The children we visit in the classroom can best understand the concept of opposites through sensory experiences. They also learn indirectly that some things don't really have opposites. What's the opposite of pizza? What's the opposite of the smell of roses? Some things that can be physically apprehended can't easily be placed on a continuum the way fast/slow and soft/loud can. Some things are subjective.

Each of our senses is processed in its own specific location in the brain. Similarly, actions, or motor functions, such as tapping the fingers or scratching one's head, are localized in other specifically reserved areas. Both the sensory and motor areas are located in about the same places in the brains of all people. Their networks are the primary input into the brain.

What we see in the classrooms is that many children are not accustomed to noticing and thinking about the qualities of sound. Parents and teachers may more often help them learn to interpret the world by calling attention to the sight and taste and feel of things. Whatever the reason, we see that quite a few children stretch and strain to identify higher and lower sounds.

Very often they confuse high and low with loud and soft. In other words, they confuse high or low pitch with high or low volume. I imagine this comes from the colloquial expression that turning the television "up" makes it louder, and turning it "down" makes it softer.

What we also see in the classroom is that while most children can easily discern whether a sound is louder or softer than another sound, and whether a passage is played faster or slower than another passage, many children have difficulty knowing whether one note is higher or lower than another. They have difficulty hearing and naming differences in pitch. This brings us to the subject of absolute pitch, or perfect pitch, a much misunderstood musical phenomenon.

It's a myth that all professional musicians have perfect pitch, and a myth that people without perfect pitch are tone deaf, or "tin eared." Very few people have perfect or, as scientists refer to it, absolute pitch. In the conservatory where I studied, I remember one student who could identify all ten pitches when someone simultaneously struck keys on the piano using all the fingers of both hands. But this ability is extremely rare. Most musicians compensate for the lack of absolute pitch with a refined sense of *relative* pitch. They know immediately the interval between any two notes, so that if they know what a D sounds like, for example, they can use the sound of that note to identify any subsequent pitch they hear. As a former oboist, I am so familiar with the exact tone qualities of every note played on an oboe that I appear to have perfect pitch—but only when I hear the oboe.

Not many people have it, but there are theories that most children could be trained to have perfect pitch. I remember a woman in Schaerbeek, a suburb of Brussels, who taught solfège, or sight singing, at a music school. She drilled her grade school students in naming notes several times a day, over a period of several weeks. Amazingly, most of these children did acquire absolute pitch, although no follow-up has been done to determine whether the effect was long lasting.

Perfect pitch is a mysterious phenomenon, seemingly a miracle when it is discovered in a child. I would define it as the infallible knowledge of the name of any note one hears. As such, it is the memory of the pitch and its name. And it most often develops after early exposure to music. There are many theories as to its origin.[33] Some researchers suggest that most humans have the innate abilities required for absolute pitch but that the window for its acquisition closes sometime in childhood. I like to believe that if the association of the name of the note with the hearing of that pitch is made and reinforced before the window closes, a child would retain the gift of absolute pitch. So here is another skill that could be taught early, and another question for scientists to answer.

Like so much else we learn, the ability to distinguish pitch is a matter of learning the names of things, and remembering them when we encounter those things in different contexts. Most people could have better pitch discrimination if paying attention to pitch were a natural part of their infancy and early childhood. Just as parents talk to their babies about the yellow rubber ducky, the green grass, the blue sky, the white sweater, and the pink blanket, they could name the pitch of various sounds in the child's environment.

I recall many times in choir, when a piece went too high for the poor sopranos, the director would transpose it down a whole tone, and we would sing lower, without paying the slightest attention to the fact that we were singing the piece in F while we were reading it in G. (By the way, this drove one classmate crazy because she had absolute pitch and just could not sing F when it was written G.)

HOW I WONDER WHAT YOU ARE

Much of the music the quintet plays in the classrooms is new to the children, but there's one tune virtually all of them recognize. When the musicians launch into this piece, without naming it, the children spontaneously start singing along: "Twinkle, twinkle, little star..." Chances are most of them know a couple of other versions of the same tune: "A, B, C, D, E, F, G" and "Baa baa black sheep." Yet the majority of them have trouble answering questions about the melody itself.

One of the musicians will say, "I'm going to play the first two notes of 'Twinkle, Twinkle, Little Star,' and I want you to tell me if the second note is higher than the first note, or lower, or exactly the same."

Invariably, the kids' responses are about evenly distributed among the three options. The musician will provide the correct answer, and then play the second two notes and ask the same question. The note lesson continues, two by two, through every note of the tune.

By the end of this process most of the children no longer have to struggle with the comparisons. No longer does every note cause them to wrinkle their foreheads and wonder what it is. They're getting a sense of what *higher* and *lower* mean in terms of music. They're getting a sense of relative pitch.

MUSIC THEORY FOR MOMMIES AND DADDIES

Higher or lower: What is going on here? Our primary grade children don't need to know what is going on, physically and physiologically, when they listen to a woodwind quintet, but adults who are trying to understand how listening to music might improve reading ability can benefit from knowing something about sound.

Although I can accept the idea that music grew out of humming and then song, it is sometimes hard to figure out how Beethoven's *Ode to Joy* or Bach's *The Well-Tempered Clavier* ever came into existence. In fact, many of the rules that govern musical composition are actually based on some of the properties of sound; others grow out of the way human beings perceive sound; and still others grow from our innate expectations of shape and form, tension and release.

At the simplest level, what music and speech have in common is sound. As the quintet members teach, sound is caused by *vibrations*—specifically, vibrations we can hear. When the vibrations occur at random speeds and over a wide range, what we hear is noise. When the vibrations are all of the same speed, called *frequency*, what we hear is a specific *pitch*. The frequency of the A note the oboe uses to tune the orchestra is 440 vibrations per second. Vibrations can be generated by vocal cords, a tightly stretched string, a column of air in a trumpet, or a struck object like a bell. The size of the vibration—its *amplitude*—determines how loud the sound is, just as the frequency determines its pitch.

Sound is a form of energy and, as such, is subject to the second law of thermodynamics, which tells us that energy runs down or redistributes itself over time. As a sound starts to dissipate, and also because of how the sound was generated, the string or column of air vibrates in parts, like a hoop wobbling as it loses its momentum. All these wobbles, called *overtones, harmonics,* or *partials,* follow precise laws of physics. A violin string always vibrates in its entire length, as well as in halves, thirds, quarters, fifths, sixths, and so on, of the whole length. These partial vibrations produce audible sounds. When any pitch is produced, the strongest overtones are the octave and the fifth.

The sum total of the vibrations and overtones makes up the particular sound quality, or *timbre,* of the instrument. Every instrument has a different set of wobbles. So a trumpet sounds different from a violin or a flute because of the differing amounts of one or the other overtones. The overtone series has significant implications in harmony, as well as in timbre and, as we shall see, the structure and form of music and the recognition of spoken sounds, like vowels and consonants.

DO(E), A DEER...

A musical scale is a collection of successive pitches. A scale is like a staircase with predictable but unequal spaces between the steps. Different cultures prefer different scales. The seven-note scale is the one we are most familiar with, but much Asian and folk music is based on a five-note, or pentatonic, scale. The first note of a seven-note scale is known as the *tonic,* and the fifth note is called the *dominant.*

Most of the music written for the seven-note scale falls into one of two modes—the *major* or the *minor*. To most people, the major mode sounds happy, and the minor mode feels sad or introspective. Children in the primary grades easily relate to the emotional tone of music, and they most frequently use the words *happy* and *sad* to describe how particular pieces of music make them feel. As adults, we might expand the vocabulary to say that a piece in the major mode sounds "triumphant," "ecstatic," "jubilant," or "merry," and that a piece in the minor mode sounds "eerie," "regretful," "reflective," or "dark."

Surprisingly, the huge differences in emotional feel between major and minor are accomplished through changing only one note in the scale. If a big step separates the second note from the third, we get the major mode. If the interval between these two notes is only a small step, we have the minor mode. It takes only one note to change the mode and the mood and the meaning of music.

As you go up a scale on a piano, for example, from middle C to the eighth white key to the right, that note will sound more like the initial C than a brand-new note. It is also called a C, and it's the first note of the next octave. Each *octave* vibrates at twice the speed of the one to the left of it on the piano keyboard. The higher the frequency, the higher the sound we hear. With its 52 white keys, then, the piano can play more than seven octaves. Our human ears can hear a wider range of pitches than are represented on a piano keyboard. We are able to hear sounds both higher and lower than can be produced by a piano.

NOW HEAR THIS!

Well, what *does* this have to do with the price of eggs? you may be asking.

The *overtone series* gives us not only a variety of tone qualities; overtones are also what distinguish one *vowel* sound from another. The "ah" sound of the letter *a* has fewer overtones than the "ee" sound of *e* or *i*. Frequency is not just something that distinguishes a soprano from a baritone. Frequency also makes the difference between an *s* and an *r*. As Robert Jourdain explains, consonants are essentially dividers that allow our brains to distinguish one vowel from another.[34]

All people have their own unique patterns of overtones, caused by their particular physique, coupled with their learned speech patterns and accents. The ability to recognize the voice of a long-lost friend over the telephone stems from the ear and brain's ability to process that person's particular overtone structure.

So when children learn to distinguish the sound of the bassoon and not mix it up with the clarinet, they are teaching their ears to pay attention to small differences in sounds. When they can judge that the second note in "Twinkle, Twinkle, Little Star" is neither higher nor lower than the first note, they are doing the same thing they need to do to distinguish the smallest units of speech. The ability to hear and manipulate those building blocks of spoken language called phonemes is essential to learning to decode written language. Awareness of phonemes is a requirement for learning to read.

PLAYING IN SPACE-TIME

Music is all about time and space.

Rhythm is measured (and proportioned) time; melody and harmony are spatial representations using pitch (vibrations or frequency). All the permutations of these take place in a physical space (the concert hall), which in turn affects these elements inasmuch as it reflects them accurately, or adds and subtracts elements of its own accord.

Some years ago the Winston-Salem Symphony performed part of one of our concerts in the gothic splendor of Duke Chapel at Duke University. In our only rehearsal, we started playing Mendelssohn's *Italian* Symphony just the way we had played it earlier in the week, at our home concert hall, the Stevens Center in Winston-Salem. In the new location, with its stone interior and very high ceilings, the echoes were so long that the musicians spontaneously stopped playing and burst out laughing. To make the music sound good in that reverberant space, we had to play it both slower and softer.

Similarly, conductors often know or find out that they can't perform a symphony at the same tempo in Italy that they might take in Germany. For the most part, German audiences like some time to delve into nuances, while Italian audiences prefer a livelier pace.

There is no universally correct tempo. How you manage the time depends on the space. That is one of the reasons recorded music cannot match the live thing, and is always something of a compromise. It can't be adjusted to the size of the room or the hour of the day.

For there to be any music, something has to vibrate—in space, over time—and these vibrations must be transmitted from the instruments to the listener through the air. Music's loudness or softness has to do with the size of the vibration—the pitch and dynamics, therefore, belong to a space continuum. How long the music is and how fast are measured in time. And articulation and tempo are grounded in time.

As Bob Campbell tells the kids, opposites are hard to define. It comes as a surprise to find out that things that appear to be diametrically different may in fact be closely related. Things we assume to be separate and distinct may turn out to be made of the same stuff and may be located on the same continuum. Like time and space? Like language and music?

CHAPTER 10

As Time Goes By

As soon as the opening piece ends, Eileen walks to the front of the class and, without a word, claps her hands twice and gestures expectantly. The children clap twice in response. Then Eileen claps four times. The children give her back an exact copy. As Eileen claps out faster and more complicated rhythms, the kids' enjoyment builds and they continue to match the patterns accurately.

"What were we clapping?" she asks.

"The beat," a little girl guesses.

"It has to do with a beat, but we were clapping rhythms," Eileen says. "Where do we find rhythms?"

"In cars."

"Where in cars?"

The children identify the persistent tone of the car alarm, the sweep of the windshield wipers, and the clicking of the turn signal as rhythms.

"Our bodies have rhythms, too," Eileen says. "What are some of those?"

Heartbeat and pulse are the first body rhythms the children think of. Then blinking. Then walking and talking. Then they go a bit further afield.

"Intestines," one child ventures.

"Beating on your desk, when you're not paying attention to the teacher," one boy says. "I do that a lot."

"Snoring," another boy says.

"Yes, I remember that my dad was a very rhythmic snorer," Eileen says.

"*Rhythm* is on top of the *beat,*" she continues. "Debi is going to play the beat. That's the part underneath that's really steady. While Debi is playing the beat, I'm going to clap some rhythms on top."

Debi raises her flute and plays a steady one-two-three-four, one-two-three-four on a repeated note. As she plays, Eileen claps out several different rhythmic patterns.

"See how the rhythm is different from the beat?" Eileen says. "The two things work together. Who wants to try clapping some rhythms?"

A boy volunteers and comes to the front. He claps ONE-TWO, one-two-THREE, ONE-TWO, one-two-THREE while Debi plays the steady four-count beat over and over.

Several more kids raise their hands and get a chance to come up and clap rhythms. Each one tries different patterns, and they all seem to find this exercise easy and pleasurable. One confident little girl claps a complex syncopated rhythm clearly derived from exposure to the world of soul, jazz, and pop music.

Then Bob stands and says, "When we play together, we try to keep a steady beat, and we have to have some way to write rhythms down, so we can stay on track."

Bob writes the word *rhythm* on the board and comments on what a strange word it is, with no real vowels, only the *y* acting like a vowel.

"We divide our music into measures, and show it with lines on each side to separate the measures," he says, drawing two parallel vertical lines on the board. "We put the rhythm inside the lines."

Bob shows them what a whole note looks like and explains that a whole note fills the measure, just like a whole gallon fills the jug and a whole pizza fills the box.

Seizing on the nearest object in sight, he says, "If I cut this table in two equal parts, what would happen?"

"I'd be in a lot of trouble," he answers himself. "But what would we call the two pieces?"

"Halves?" one boy ventures.

"Halves. That's right. Like a candy bar or anything you split in two pieces, each piece is a half. So to fill up a whole measure, how many half notes do you think we'd need?"

"Two."

"Yes, two. One half plus one half equals a whole. This is what a half note looks like." And he draws two circles with stems on them, which distinguish them from whole notes.

As Debi plays the steady four-count measure, Bob leads the children in clapping one whole note to fill a measure, then two half notes to fill a measure, clapping the half notes on the first and third beats.

Bob helps the children think of other things that can be divided into halves. Football games. Peanut butter sandwiches. The earth, by its equator.

"Your beard," a little girl says, and comes forward to show how she would divide Bob's beard into halves.

Bob then draws four quarter notes, filling in the circles and attaching stems.

"How many quarter notes would fill a measure?" he asks.

Some of the kids clearly find this a tough question, but a few raise their hands. A little girl responds, "Four quarter notes."

"Right, four quarters. What else comes in quarters?"

"A dollar," one child says.

"A Quarter Pounder," suggests another.

"Why do they call it a Quarter Pounder, do you think?" Bob asks.

This is a puzzlement to the children. They haven't thought about it before.

"It's a quarter of a pound of meat, because a whole pound is way too much for a hamburger. Even a quarter pound makes a pretty big hamburger."

Bob draws measures on the board, some containing whole notes, some with two half notes, and some with four quarter notes.

Debi picks up her flute again, and plays the beat as the children read each measure and clap them out in sequence.

For most, it's their first lesson in "reading" music, and they do it perfectly.

"Give yourselves a hand," says Bob, and the kids clap enthusiastically, with no concern at all for rhythm.

Fractions are when arithmetic starts to get hard for some children, and in our first year at Bolton the second-grade teachers asked for some help from the quintet in getting across this concept. Fractions are thoroughly familiar to musicians, but we call them rhythm. This is one of the more obvious examples of an academic subject that music can help teach, and one where it's easy to see immediate results.

"You can see the lightbulbs go on in their heads," says Eileen Young, who has been the quintet's clarinetist from the project's beginning. "We're targeting the kids who didn't understand the concept until we introduced it through music. I remember being one of those kids."

After this first lesson in fractions, which is easy to grasp and builds the children's confidence, the quintet moves on to advanced lessons, in which they introduce eighth and sixteenth notes, and in which the children learn all the combinations of fractions that can make a whole. The quintet uses the example of gallons, half gallons, quarts, and pints to help the children understand which units can be combined to fill the whole, and which would create an overflow. They talk about slices of pizza, and coins, and quarters of basketball games. Then they ask the children to write lines of music, filling each measure with a combination of parts that would add up to a whole measure. The musicians then play each child's composition for the whole class.

A class in rhythm and fractions will often conclude with one of the musicians playing a melody while another quintet member claps the rhythm. The kids then listen to the complete piece of music played by the whole quintet, listening especially for the rhythm.

Rhythm is also the basis of the lessons the quintet teaches in poetry. A popular lesson with second graders makes use of Jack Prelutsky's funny story poem, "The Turkey Shot Out of the Oven."[35]

*The **tur**key shot **out** of the **o**ven*
*and **rock**eted **in**to the **air***
*it **knock**ed every **plate** off the **ta**ble*
*and **part**ly de**mol**ished a **chair**.*

It ricocheted into a corner
and burst with a deafening boom,
then splattered all over the kitchen,
completely obscuring the room.

It stuck to the walls and the windows,
it totally coated the floor,
there was turkey attached to the ceiling,
where there'd never been turkey before.

It blanketed every appliance,
It smeared every saucer and bowl,
there wasn't a way I could stop it,
that turkey was out of control.

I scraped and I scrubbed with displeasure
and thought with chagrin as I mopped,
that I'd never again stuff a turkey
with popcorn that hadn't been popped.

The class is divided into five groups—one for each of the poem's verses. Each member of the quintet works with one of the groups, first having them clap the rhythm, then teaching them one verse of the poem. The triple meter of the first verse above is highlighted in boldface, but by lengthening the accented syllable, a hip-hop version can be created. Independently, the verses don't make much sense, but when the groups recite each verse in order, they add up to a good story with a surprise ending. It usually takes part of three lessons for the kids to memorize the verses. Sometimes the "performance" lacks precision and polish. Then the groups break up for more practice. It's important that each group has its lines down pat, to make the whole piece work.

Robert Franz, the first coordinator of the residency project, considered this lesson the perfect analogy for the quintet itself, in that it shows the way the members work on their own parts of the music separately and then bring them together in a surprising and satisfying whole.

The vocabulary in this poem is well above second-grade reading level, and it's likely that many of the children have never before encountered *chagrin, ricocheted,* and *obscuring*. But the combination of rhythm and rhyme makes it easy for them to learn and remember the verses.

Most children today are not accustomed to committing poems to memory, but the appeal of highly rhythmic speech continues. Today almost all children are exposed to rap and find it attractive. Similarly, poetry slams are popular with some older students and adults, as is African drumming. Rhythm is the common element of music and poetry, and even as we see traditional ways of experiencing music and poems fading away, rhythm asserts itself in newer forms, like rap. Rhythmic expression appears to be eternally appealing, if not somehow essential, to human beings.

Clapping rhythms seems to be completely natural and easy for most children to do. For years the Winston-Salem Symphony played a series of concerts for children from about age three to nine, and the children spontaneously clapped along with the orchestra during many different kinds of music.

As we have seen, music has both pitch (melody and harmony) and rhythm. Rhythm is measured time, time divided into interesting packets, some predictable, some not. You can see how fundamental it is to music by clapping the rhythm of a familiar song to a friend and asking the friend to name that tune. Many people can identify a song from its rhythm alone in a couple of seconds. When I work with young children I will often clap for them the rhythm of some well-known song, like "Happy Birthday," from the stage, without telling them the name of the song, and ask them to start clapping along with me as soon as they recognize it. Again, this is a simple exercise for them—and I notice they find it easier to recognize a tune by its rhythm than by its pitches alone. That leads me to conclude that rhythm is more basic to music than pitch.

We know rhythm is ancient and fundamental to music, as it is to poetry and dance. But there are still things we don't understand about rhythm and its role in enabling people to perform a whole range of physical and mental tasks.

I learned of an interesting recent development from Don Flow, a well-known businessman and community leader. I know him as a music lover and sponsor of the Winston-Salem Symphony, who traces his enjoyment and appreciation of music back to his own elementary school days, when his class was taken to hear the symphony perform.

One of Don's college classmates had a son with a physical disorder that resulted in the boy's inability to walk. The father had read that learning to do something rhythmic sometimes helps people improve their timing in other activities, a kind of cross-training effect. Following up on this, he talked to a click track expert whose background was in Motown music. Click tracks are what recording artists use when not all the musicians who will be heard on a final recording or film are available to record the music at the same time. Having each individual artist keep time with the signals clicked into headphones ensures that they will all "play together" in the finished product.

The Motown expert went on to develop a computerized interactive metronome to improve people's timing. It involved clapping on the beat of a slow, regular, computer-generated signal—something like a mechanized version of what Eileen and Debi were doing with our third graders. A feedback mechanism let the user know if he was ahead of or behind the beat. The goal was to clap precisely on the beat.

Don Flow was a backer of the early research on this metronome, and he discovered that it could also be used to diagnose talents and deficiencies of timing. He tried it out with all the mechanics at his Cadillac dealership. It turned out that the mechanics who could best coordinate their clapping with the beeps of the metronome were the ones who were, in fact, the best and most productive workers, the men who could most efficiently diagnose and repair problems. There was one exception to this pattern. A young man who did a great job of staying in sync with the metronome was not one of the top performers on the job. In fact, he was new to the job and inexperienced. His rhythmic ability probably stemmed from his avocation: he was a drummer in a rock band.

Our musicians began teaching rhythm as a way to teach fractions, but it seems likely that these lessons contribute to improvement in other areas, including attention and reading. Rhythm is with us every minute. We walk with a steady binary rhythm: one-two, one-two. After hard exercise, we breathe in a very fast binary rhythm: one beat to inhale, one beat to exhale. We are born with the memory of the rhythm of our mother's heartbeat. And mothers often instinctively hold their babies on the left side of their bodies, to give the baby the comfort of this most familiar rhythm.

A *beat* is like a heartbeat, a regular pulse underlying all Western music. A beat is like a clock ticking—it is evenly measured time. *Rhythm* divides beats, or sometimes combines beats, into recognizable patterns. The rhythm is what makes a piece of music a waltz, fox-trot, tango, or cha-cha. The rhythm is also what gets our bodies moving and involved with a piece of music.

We have seen that the scale or mode influences our emotional impression of a piece of music. Rhythm also touches our emotions. A quick rhythm feels positive and optimistic. In fact, we call that state of mind *upbeat*. A slow, heavy, repeated rhythm can have the quality of a funeral march or dirge. Many activities of our lives have their associated rhythms.

We've heard that "timing is everything." A truer statement might be that timing is critical to many of our mental and physical activities. Some people discover for themselves that adding a rhythmic component to certain tasks helps them do better. For instance, I have a niece who is an Olympic-caliber volleyball player. She's found that she can do her schoolwork better if she uses one hand to keep a ball steadily bouncing while she studies. I know a man in his 70s who likes to climb mountains, and who gets to the destination ahead of many much younger people. That may have something to do with the fact that he is listening to marches on headphones when he makes these treks. Many students report that they can concentrate on their homework better with music playing in the background.

Besides this informal, anecdotal kind of information, a body of research is now showing us how timing and rhythm explicitly relate to cognitive functions, such as attention and academic achievement, as well as to motor functions, such as planning, sequencing, and coordinating of movement.

Dr. Stephen M. Rao and his colleagues at the Medical College of Wisconsin in Milwaukee have recently identified specific areas of the brain that govern the body's sense of timing.[36] Dr. Rao speculates that attention deficit hyperactivity disorder (ADHD) may involve a disorder of time perception.

A good deal of research links rhythm to another learning disorder, dyslexia. Severe dyslexia is now understood to be a learning disability that has a neurological origin. While deficits in phonemic awareness are at the core of many kinds of profound dyslexia, other kinds of dyslexia have their origins in time and fluency processing, a second core deficit, which explains why some children with dyslexia don't respond completely to phonological-awareness treatment.[37] The third core deficit of the cluster that defines dyslexia is the inability to retrieve the name of an object or person.

Scientists at University College London have found that children with dyslexia have difficulty detecting the beat in sounds with strong rhythm.[38] They also found that children who are exceptionally good readers for their age excel at spotting rhythms. The researchers concluded that awareness of beats may influence the way young children internalize speech patterns, which may then affect their ability to read and write.

People with dyslexia, whether children or adults, can learn to read with proper instruction. The Orton-Gillingham approach has been widely used for some 50 years and is the source and inspiration of many proprietary reading instruction programs. It is named for Samuel Torrey Orton, a neuropsychiatrist who was a pioneer in the study of dyslexia in the 1920s, and Anna Gillingham, an educator and psychologist who worked with Dr. Orton.

Dr. Orton's wife, June, carried on his work, coming to Winston-Salem after his death to conduct an extensive practice with children with learning disabilities. A decade after her death, a research team at Wake Forest University headed by Dr. Frank Wood began following the children she had served—who by then had become adults with their own children. Subsequent studies on these Orton "alumni" have confirmed not only that dyslexia persists into adulthood, with significant impact on life adjustment, but also that there is a strong genetic component to dyslexia.[39] Consequently, the disorder often gets "passed on" to subsequent generations—apparently as a largely dominant trait,[40] meaning that if either parent is dyslexic, then any child has a 50 percent chance of being dyslexic.

Many of today's reading experts, including Barbara Wilson, director of Wilson Language Training and developer of the Wilson Reading System, have expanded on the work of Orton and Gillingham. The Wilson system uses sound tapping in the earliest stages of reading instruction. That is, students tap fingers against the thumb as they sound out phonemes, and drag their thumb over the fingers as they blend the sounds together. In this way, they break down words into their component phonemes and then blend the phonemes together to make words. Sound tapping is part of a multisensory approach in which seeing, hearing, and touching are all employed to reinforce the complex skills of reading.

Similarly, all the lessons of the quintet use more than one sense to convey, reinforce, and lock in learning. It's easy to see that sight and sound and movement all together make learning more engaging for kids who are still at the squirmy stage. It's easy to see that, when musicians are in the classroom, rhythm has a big presence. Looking at the kids, one can't help noticing that it takes no special effort for them to "get" rhythm. They've got rhythm, and rhythm has a particular power to focus their attention.

CHAPTER 11

Is Music a Reading Teacher?

Emma Hutchens, born in 1986, is a senior at Assumption High School in Louisville, Kentucky. Until she was in fifth grade, her family lived in Winston-Salem and Emma attended Bolton Elementary School. Some of her happiest memories of the school are of visits from the woodwind quintet.

"Everybody in the class looked forward to them coming," she said recently. "It was something out of the ordinary. They presented a different subject in a different way, and they were not like authority figures. They were more like friends."

The most memorable occasion for her was an off-campus evening concert by the quintet at the end of the school year, which many Bolton children attended with their parents. Emma was chosen to present a bouquet to Robert Franz, an honor that still gives her pleasure to think about.

Nobody ever told the children why the musicians were coming to their class, and Emma has always assumed it was to instill appreciation of classical music. It did that for her, and she uses classical recordings regularly to relax and focus her mind when she studies.

That might have happened whether or not musicians had visited Mrs. Caesar's third-grade classroom. Emma heard classical music at home, as well as over the public announcement system at school. She also remembers seeing the quintet rehearsing in the halls.

She doesn't play an instrument herself, and she is not involved in any extracurricular musical activities. She's a serious student, who hopes to become a lawyer working for social justice.

Strange as it seems to her, the musicians who were such a novelty in the classroom, and who she thought were there to teach such things as the care and maintenance of the oboe, helped her become the excellent student she is.

"The greatest benefit?" she said. "I know that that experience developed my reading skills, and my learning skills."

MEASURES OF SUCCESS

The improved reading and arithmetic scores at Bolton Elementary School were impressive, but it was impossible to assert that the musicians in the classroom had directly contributed to the test results because the study never had a control group. More recently, in 2001, Shirley Bowles, a public school psychologist working in a neighboring county, approached the Wake Forest University School of Medicine with the idea of studying the link between music and brain development for her doctoral thesis. Dr. Wood told her about the ongoing musician residencies we had initiated, and she decided she would like to test certain hypotheses with our curriculum. This was a fortuitous development for us because the musicians-in-residence project was scheduled to move to a charter school, where a rigorous scientific study would be not only allowed but welcomed. In fact, we were looking for a researcher who could design a study that would actually measure what the residency program was accomplishing.

Shirley Bowles submitted a proposal to study whether the curriculum we had developed had any impact on phonemic awareness. She also proposed to analyze whether the classes had any impact on spatial-temporal reasoning ability, using a test, called the Audio-Visual Integration test (AVI),[41] devised by Dr. Gurmal Rattan of Indiana University of Pennsylvania.

Shirley tested her hypotheses at the newly established Arts Based Elementary School, affectionately known as ABES, in the fall of 2002. The same woodwind quintet and the same curriculum used at Bolton were used at ABES, but this school was not much like Bolton, where we had run the program from the winter of 1995 to the spring of 2002. ABES, chartered by the State of North Carolina, receives funds from the state based on enrollment, at the same per-student rate as other public schools. Teachers are subject to the same certification as in the regular public schools. But instead of being governed by the publicly elected board of education, a charter school is governed by a board of directors, much like a private school or a typical nonprofit organization.

As a charter school, ABES receives no state funds for building and grounds. I was on the founding board of directors of the school, and finding appropriate and affordable space was a major challenge for us and actually delayed the opening of the school by a year. When the school finally opened, it was housed in Atkins Middle School, in five adjoining, formerly empty classrooms. In the fall of 2002, 66 children were enrolled in kindergarten through third grade. In the fall of 2003, enrollment grew to 175; more class sections and a fourth-grade class were added. By then the school had a permanent home of its own, in a renovated factory.

When Shirley did her research, which included all the children in the Arts Based Elementary School the first year it was open, half the children enrolled were Caucasian, one-third were African-American, and one-sixth were Hispanic. A full 20 percent had been diagnosed with learning disadvantages and disabilities. Almost one-quarter were classified as indigent. Our eight years at Bolton had strongly suggested that ethnicity and socioeconomic levels were not important factors in our results. The improvement rate had been the same across the races. In both schools, also, the class size was relatively small, under 20 students.

At ABES, the arts are generously infused into the basic curriculum; the premise is that the arts are inherently excellent teachers of many skills and habits of mind. The musicians-in-residence program clearly fit in with this vision.

After the initial workshop with the classroom teachers, the students in kindergarten through third grade were randomly and evenly divided into test and control groups. They were not screened for gender, race, or handedness. At the beginning of the school year, all students were evaluated by means of three tests. One was the Broad Reading sections of the Woodcock-Johnson III Tests of Achievement, which test letter-word identification, reading fluency, and passage comprehension. Another was the Predictive Assessment of Reading (PAR), a simple and elegant test for measuring phonemic awareness as well as fluency and name recall, devised by the section of Neuropsychology at Wake Forest University. And the third was the Auditory-Visual Integration test (AVI), the previously mentioned computerized test that measures spatial-temporal tasks. The children were evaluated again, using the same tests, at the end of the term, before the Christmas break.

During the course of the study, the test group had 18 half-hour lessons with the quintet over a four-month period, using the curriculum described in this book. The control group took chess lessons during the same period. Despite the fact that the study covered only one semester at ABES, compared with three years at Bolton after which the children were tested, the results were significant. Both phonemic awareness and spatial-temporal abilities improved significantly at all grade levels, regardless of gender and ethnicity.[42] And phonemic awareness, as we've seen, is the first essential of learning to read. Shirley's study indicates that children in early elementary grades who have all the same opportunities as their peers develop stronger building blocks for reading with the addition of nine hours of special music instruction.

Imagine if musicians and music could really have an effect on the illiteracy crisis in our country! That would be a discovery of tremendous significance, a giant step in solving one of the most stubborn and serious problems we face as a society.

ILLITERACY— A NATIONAL AND A PERSONAL TRAGEDY

According to the 2003 National Assessment of Educational Progress report, approximately 40 percent of students across the United States cannot read at a basic level. By fourth grade, when we expect children to read well enough to learn a variety of academic subjects, 37 percent are unable to read and understand a simple paragraph. For low-income and minority fourth graders, the rate climbs closer to 60 percent. Despite all the attention this problem has received, a downward trend has proved to be tenacious. The groups of students who perform at the average level have made no progress in the past ten years. The lowest-performing readers have actually lost ground.[43]

When learning to read is the focus of the first four years of school, from kindergarten through third grade, why are so many children failing to learn? Experienced teachers often comment that children today are harder to teach than earlier generations. Income levels aside, more children today appear to have trouble paying attention, listening to the point of view of others, and focusing on a task. Teachers notice that they have smaller vocabularies and a greater tendency to speak in vague terms such as *like* and *you know*. Many kids today have trouble hearing differences between sounds in words, and trouble understanding long or complex sentences. They are not tuned in to the differences between colloquial speech and written English. And in both speaking and writing, they have diminished ability to arrange facts in an orderly sequence.[44]

Children who fail to master reading in the early grades rarely learn to read later in life. In the United States now, an estimated 43 million adults are functionally illiterate.[45] Only an estimated 17 percent of adults have the reading proficiency and the education to work effectively in our technologically complex world.

Illiteracy is now considered a health issue by the National Institutes of Health (NIH). Statistics indicate that *difficulty in reading is a major risk factor for suicidal behavior.*[46]

For this shocking state of affairs, there is plenty of blame to go around.

Parents are blamed, for not providing the structure and discipline that were more common in earlier generations. The schools are blamed for methods of teaching reading that may actually be causing reading failure. Many facets of modern life and forces in contemporary society are cited as contributing factors: television; the testing movement; an upswing in attention deficit hyperactivity disorder; an epidemic of behavioral, emotional, and learning disorders; sugar; secular humanism; mainstreaming; mothers' working; an emphasis on visual information; an erosion of traditional values; entrenched poverty; and environmental poisons. *Why Johnny Can't Read,* by Rudolph Flesch, was published in 1955, and the reasons have been piling up ever since.[47]

Separated from all the cultural freight, the fact remains that some children learn to read fairly easily, and other children have tremendous difficulty. Most primary grade classrooms contain both kinds. Even the same family may include both kinds. From the beginning, the Bolton project has strongly suggested that some children who haven't been able to learn to read with customary methods of instruction seem to benefit from music. To understand why that might be, we need to look at commonalities between language and music.

Human infants learn to talk and understand the syntax of the language spoken around them with no special effort or instruction. We're hardwired for language. Steven Pinker, Johnstone Family Professor in the Department of Psychology at Harvard University and author of *The Language Instinct,* points out that when a group of children are thrown together without a language they can use, they will invent one. What they won't do is invent a written language. In Pinker's words, "A group of children is no more likely to invent an alphabet than it is to invent the internal combustion engine."[48]

Dealing with print is not something we humans are born to do. Reading and writing are complex processes that don't come spontaneously. Just as we have to be taught to swim or to drive a car, we need some help in learning to read and write. We need to be taught that the words on the page are an attempt to set down the sounds of our spoken language. We need to know that the letters and the words in books are a code, and we need to be able to crack the code.

Spoken language is ancient in terms of the evolution of our species, and probably developed about the time our ancestors learned to walk upright. Written language, however, is relatively new. The first writing system began to be developed around 3500 B.C. by the Sumerians, to keep records of agricultural products. Somewhat later the Egyptians, the Chinese, and the Mayans of the Yucatán jungle followed a similar process for similar reasons as they tried to encode information that others would be able to decipher. All the early attempts to turn speech into written symbols focused on the meaning of the words and used abstract symbols to stand in for word meanings. Those first attempts failed because the thousands of symbols they required defied human memory. Human beings can remember thousands of words, but we are incapable of learning and remembering thousands of abstract symbols to represent them. Written language only worked when it began to use otherwise meaningless symbols to stand for particular sounds that were used over and over, in many combinations, in thousands of words.[49]

PHONEMES ARE FUNDAMENTAL

Different languages have different numbers of basic sounds, called phonemes, that are necessary for speech. Using our earlier example, the word *cat* is made up of three distinct phonemes: /k/, /a/, and /t/. American English uses about 44 phonemes, and several African languages use more than 200. But no language has an infinite number of sounds, or even thousands of sounds. When the marks on the page only designate sounds, not meaning, you have a manageable number of symbols. When the symbols stand for sounds, the number can be small enough for people to remember easily.

The collection of symbols used to represent the sounds of a language comprises its alphabet plus a small number of diacritical marks (*é, è, â, ç, ü, ñ*). Different languages have different numbers of letters in their alphabets: English has 26 letters; Russian has 33; Italian has only 21. It is important to note—and for children to learn—that we don't always have a separate letter for each phoneme. The same letter may have different sounds, or a combination of letters may be used to signify a distinct sound.

English is an especially difficult language to learn to read and write. We have numerous ways of pronouncing various letters and combinations of letters, which makes it hard to sound out many words and hard to know how to spell words. An often-cited example is "ghoti," which could reasonably be pronounced "fish"—the "gh" sounded as in *tough,* the o as in *women,* and the ti as in *nation.* While Italian- and Spanish-speaking children typically can learn to read and write in a year of school, British and American children take at least two years.

MUSIC AND LANGUAGE—EVOLUTIONARY SIBLINGS

Music, like language, seems to come naturally to human beings. As we have seen, some of the earliest artifacts unearthed show that our prehistoric ancestors made music. And like language, music existed for eons before anyone tried to transcribe the sounds into symbols on a page. The ancient Sumerians, Greeks, and Hebrews developed forms of musical notation, some of which are forerunners of our current notation systems. In the ancient world, however, music, like its sister arts, poetry and storytelling, was for the most part an oral and aural tradition, handed down from one singer or player to another.

In our processing of them, speech and music share many characteristics. Both are transmitted by sound waves, captured by the ear, and then converted into neural impulses. In the case of song, both are produced by the vocal cords and mouth. While images of brain function seem to indicate that most of the processing of both speech and music takes place in the same general regions of the brain, some parts of the brain are dedicated primarily to processing language, and others primarily to processing music.

Language and music share other important characteristics in the brain. The separate elements of music—such as pitch, rhythm, and emotion—are processed in different parts of the brain and reassembled to make what we experience as music. Similarly, language is broken up into the perception and processing of phonemes and meaning and comprehension. Music and language both rely on the perception and processing of assembled units with temporal and tonal features that are associated with unique symbols—notes in the case of music, letters in the case of language. Both music and language are multisensory.

That language and music share so many features and yet remain distinct functions raises the question of cross-training. As shown in Shirley Bowles's study, learning to listen acutely to music probably enhances at least one skill necessary to reading: phonemic awareness. Perhaps it may strengthen others, such as comprehension and fluency. Answering these questions will be the focus of future studies, which I am helping to design with the section of Neuropsychology at Wake Forest University Medical School.

READING—A COMPLEX SET OF SKILLS

Many people reading this book may barely remember how they learned to read because it came so easily to them that any type of instruction would have worked. There is now a sizable body of scientific research on teaching and learning reading. Research-based reading instruction, such as the methods based on the Orton-Gillingham approach, is especially suited to children who have difficulty learning to read. The federal government, through the Reading First initiative, is putting a major emphasis on these types of programs, and channeling resources into them.

What the research shows is that children need to become proficient in five specific areas in order to become good readers. These components must be explicitly taught to children who have difficulty learning to read:

Phonemic awareness is first. It refers to the ability to hear, identify, and manipulate individual sounds—or phonemes—in spoken words.

Phonics is the relationship between the letters of written language and the sounds of spoken language.

Fluency refers to the capacity to read a text accurately and quickly.

Vocabulary refers to the body of words students must know to communicate effectively.

Comprehension is the ability to understand and gain meaning from what has been read.

By most informed estimates, some 25 percent of children run into difficulty in learning to read, for a variety of reasons.

The majority of them qualify for the diagnosis of dyslexia. Dyslexia, which literally means "poor reading," used to be thought of as a visual disorder. It is now understood as a difference or deficiency in language processing. Specifically, individuals with dyslexia have trouble processing phonemes. Some experts estimate that as much as 20 percent of the American population is afflicted with varying degrees of reading disorders,[50] which cannot be cured but can be compensated for. Dyslexia can now be diagnosed easily in the early grades, using such tests as the Predictive Assessment of Reading. The PAR test simply asks a person to separate words into their component sounds, and to leave out one sound.

The test is so simple that I've tried it with many friends. One is a visionary man who spent almost 30 years developing an art museum that is acclaimed for its vibrant educational and outreach programs. He's a fine painter and a decent musician. I would also describe him as scholarly and erudite, a fountain of knowledge on all kinds of things. I said to him one day, "Say 'cat' without the first sound." He paused for a moment and appeared to be wrestling with the task. "I can't do that," he finally said. "Well, say 'cat' without the last sound," I urged him. "I can't do that either," he said.

This man, in his 60s now, is profoundly dyslexic. He claims that his dyslexia also hampers his ability to read music. However, he's an example of the truth that dyslexic doesn't mean dumb. The processing of phonemes has no relationship at all to intelligence. George Patton and William Butler Yeats were both dyslexic. My friend also demonstrates that dyslexia is not a stage that is outgrown but a condition that persists through life. Two of his children are also profoundly dyslexic, illustrating what science has recently confirmed, that severe dyslexia is usually transmitted genetically.

Dyslexia, let me stress, is a problem with *reading,* not with understanding language or being able to play music. If music can improve phonemic awareness, as we have shown, there is hope that perhaps music may become a coping tool at least for youngsters with mild to moderate dyslexia.

All children need to be able to hear phonemes in order to read, and extra instruction and practice in recognizing phonemes can help children with dyslexia. We know that every lesson from the woodwind quintet trains children to notice even slight variations in sound, and to identify these variations. Yet when we pay close attention to what the quintet does in the classroom, it is also clear that every lesson reinforces other components of becoming a good reader.

Every lesson is also a phonics lesson, as the musicians write words such as *piccolo* and *forte* and *bassoon* on the board and have the children sound them out. Many lessons teach fluency, as when the children graph a line of music and then sing it wordlessly. All lessons expand the children's vocabulary, as the musicians provide richly elaborated information about themselves, their instruments, and musical concepts, as well as basic academic subjects. And comprehension is absolutely integral to the entire curriculum: the musicians constantly question the kids to verify their understanding of the concepts they are being taught.

Emma Hutchens, who was in the first group of children to experience the musician residency, said the classroom teachers she had at Bolton were "great," but the musicians' style was entirely different.

"They didn't lecture. They said we were going to learn together in a group and discuss what we had learned," she said. "This was not the norm at Bolton. Most teachers would just teach something once. The quintet members were very focused on helping us learn, and they would present information in a lot of different ways. We did a lot of reading out loud, and I learned from them that there are different ways of learning things."

The musicians are doing intuitively what reading experts—and great teachers everywhere—know is the most effective way to teach reading. They make certain that all the skills reinforce each other. The skills are separate, but when they are pulled too far apart, they become dry and dull. It's hard to learn vocabulary from lists of vocabulary words. It's boring to spend a half hour sounding out fragments of words. It's discouraging to try to become fluent in prolonged sessions that point up your obvious lack of fluency. Kids are able to maintain interest and energy when they work on several skills simultaneously.

Linear thinking is not particularly interesting to the musicians. Their lessons are varied and full, yet they also have clarity and cohesion. As artists, musicians seem to have a good sense of what "hangs together," and so we should not be surprised that any reading component they teach automatically calls up the whole set. If the lesson is *staccato,* the musicians will quite naturally say it expressively, write it, define it, graph it, use it in a sentence, jab the air with their arms, play it on their instruments, ask for examples and analogies—and review it the next week.

Even if they are not explicitly teaching reading, they are certainly providing a lot of information about ways to approach the written word.

CHAPTER 12

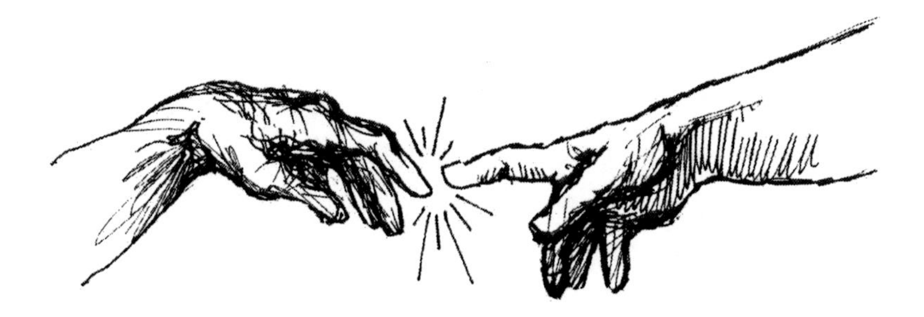

Listening to Learn

The third graders are lying on their stomachs, looking up at Debi, Cara, Kendall, Eileen, and Bob and concentrating on the beautifully haunting music the quintet is playing for them. It's the musicians' sixth visit to their class. By now the children are good at identifying whether what they are hearing is loud or soft, fast or slow, high or low, and legato or staccato. Today, Bob has asked them to listen for the musical opposites, and also to listen for what the music reminds them of and what it makes them feel. He's asked them to write down "describing" words as the music washes over them.

When the piece has ended, most of the children raise their hands, ready to volunteer the adjectives and images that came to their minds. Eileen writes them on the board: scared, soothing, sad, smooth and also bumpy, spooky, like somebody died, like tiptoeing.

When one boy says it was loud, Bob looks thoughtful. "Did it sound loud to you?" he asks. "Maybe it did, because soft and loud are relative. Something can sound loud to one person and soft to another. To me, it seemed we were playing softly, because I know how much louder than that all our instruments can play."

Bob tells them the piece is the Pavane by Gabriel Fauré.

"A pavane is a slow dance," Bob says. "Fauré is the name of a French composer."

Now the musicians turn to another piece of sheet music on their stands, and Bob asks the kids to notice how this piece is different from the pavane and to write down words that describe it.

This piece is dramatically different in feeling, and some of the kids laugh out loud as they hear it. When it is over, they come up with another set of descriptions: forte, fast, funny, exciting, bumpy and happy, joyful, like a clown dancing, like sailing over the waves, like the Hokey Pokey.

Eileen writes the word *polka* on the board. A polka is a dance, too, but much livelier than a pavane.

"Now turn your paper over and write four or five sentences about which piece you liked best," Bob says, moving to the board. "I'll give you some help to get started. You can write, 'The piece I liked best was the ____________.' Then you write either *polka* or *pavane*."

"What if I liked both of them best?" a boy asks.

"Then you can write what you like about both of them," Bob says.

The quintet members leave their stands and fan out among the children, ready to give individual help and encouragement to anyone who looks stumped or intimidated by the blank sheet of paper.

What's going on when an eight-year-old child searches his mind to describe Fauré's Pavane? What resources is a third-grade girl drawing on when she writes about a Shostakovich polka?

One process that is happening is a kind of comparison of this new information, the pavane or the polka, with what is already stored in the brain. "What does that make you think of?" and "What does that remind you of?" are good questions to stimulate thinking and the imagination.

You may know the process of brainstorming, which is used in many work settings to come up with innovative solutions to problems. In brainstorming, the people in a group are encouraged to call out anything that comes to mind in response to a particular question or subject. Every idea that is spontaneously offered is written down. Initially, there is no attempt to evaluate or prioritize the suggestions. The goal is to generate the largest possible number of ideas.

Creative artists of various kinds use similar techniques. They may start with a particular word or image or event and then free-associate, pulling in dozens of ideas and images that are all associated in their minds with the original thought.

Anyone can do this. You might notice a cup of coffee on your desk and quickly write down all your associations with *coffee*. After a minute or so your list might include morning, caffeine, Colombia, Costa Rica, mug, joe, "Let's have another cup of coffee, let's have another piece of pie," break, ice cream, Java, computer software, Starbuck's, beans, grinder, all-nighter, cream, latte, plantation, and so on. Your associations with *coffee* might include snatches of songs, jingles, or scenes from television commercials; memories of places where you have sipped coffee; embarrassment at remembering spilling coffee on someone or something; queasiness at the memory of drinking too much coffee; and much more.

Anyone can do this because that's the way our brains work. Hooking one thing to another is what the brain does more than anything else. Each of the estimated 100 billion nerve cells (neurons) in our brains is capable of making more than 10,000 connections with other cells.[51] Scientists believe that the number of possible connections in one brain is similar to the number of stars in the whole universe.

PRACTICE MAKES PERMANENT

A word like *coffee* is represented in our brains as a network of neurons. Any single group of neurons can participate in thousands of networks, like the "morning" network of associations, the "Costa Rica" network, the "ice cream" network, and on and on. Most of these networks were created with some sensory and/or motor input. For example, an unexpected whiff of freshly made coffee can flood you with memories. Those memories may include other sensory input, such as where you were at the moment; what sounds were in the background; whether you were hungry, cold, or alone; and many other details. Some cells in all these areas in the brain (such as the smell area, the auditory area, the somatosensory area, the "warm-fuzzy" emotional processing area, and more) link together by firing at nearly the same time. However, the neural network incorporating sensory input with hardwired bodily sensations—created by, say, the smell of coffee, the look of your kitchen, the sense of being hungry, the feeling of being cold, and the fact of being alone—is initially weak. Unless this particular constellation is repeated, the association may dissolve. But if just about every morning you come to the kitchen to the smell of fresh coffee, with your stomach growling in anticipation of eating breakfast, feeling chilly in contrast to the warm bed you just crawled out of, and alone with your own thoughts and actions—those associations will form an enduring cluster in your brain.

Striking examples of how sensory and motor input set up lasting associations comes from imaging the brains of musicians. A musician *listening* to a piece of music that he or she has learned and played in the past will activate the part of the brain that controls the *fingers*. That tells us that neural networks were created and reinforced when the musician repeatedly practiced that piece of music.

You may experience something like this if certain kinds of music virtually force you to tap your toes or move other parts of your body in time to the beat. Similarly, the thought of the smell of a sewer or a skunk may cause you to involuntarily wrinkle your nose. A picture of an iceberg may make you shiver. Information from our senses becomes an enduring part of our mental representations of songs, skunks, sewers, and snow. One of my favorite tricks in my youth was to suck a lemon in sight of band musicians, who were then unable to play because of the involuntary pucker their mouths made.

I see a variation of this in the youth competitions that I am often called on to judge. Over the years I've noticed that when a child makes a mistake or loses her place in a memorized piece, the slip almost always occurs at a place in the piece where a page in the sheet music must be turned. As the child practiced the piece over and over, she had to pause to turn the page. I suspect that what gets reinforced in the brain is the memory of stopping at these particular places. Only perfect practice makes perfect!

A SYNAPSE IS A TERRIBLE THING TO WASTE

Without any conscious effort, we constantly create new neural networks. In fact, that is as good a definition of learning as we are likely to find: the creation of neural networks.[52]

What is actually happening in the brain is that a brain cell is transmitting an electrochemical impulse, through a gap called a *synapse,* to another brain cell. To me, Michelangelo's famous *The Creation of Adam,* on the ceiling of the Sistine Chapel, vividly depicts what the transmission of an impulse from one neuron to another is like. You can almost see an electrical spark in the tiny space between God's finger and that of Adam. Scientists use the word *excitation* for the transmission of an impulse from one brain cell to another. The informal term is *firing.*

The late Donald Hebb, a Canadian neurologist, did some of the pioneering work in explaining how learning occurs. He was among the first to notice that each time a firing occurs between two particular nerve cells, the connection between them is strengthened. The Hebbian learning law is now informally expressed as "Neurons that fire together wire together."[53] The converse is also true. When neurons don't fire together repeatedly, the weak connections wither away. Hebb's findings are also articulated in a familiar bit of advice, "Use it or lose it."

Groups of neurons that tend to fire together form assemblies of cells whose activities can persist after the firing event. Each cell assembly represents a particular piece of knowledge or learning. What we call thinking is physiologically the sequential activation of sets of cell assemblies. Literally, one thing leads to another. As long as this sequential activity is moving along, what we are doing is paying attention. If we become distracted and drift off, and one thought no longer leads to the next thought, there's a disconnect.

Another important characteristic of neurons that fire together is that firing together often produces efficiency gains. This is intuitively easy to understand. When we first learn to drive a car, driving requires a great deal of effort and concentration. After we've driven for a few months or years, most of the decisions and actions required to operate the vehicle become virtually automatic. We barely have to think about turning the wheel, switching on the headlights or windshield wipers, or braking at the sight of red taillights on the car ahead.

The way we know that we know something is if we can remember it. It would be difficult to claim we had learned something if we were unable to recall it. That's what teachers are counting on when they spring quizzes and tests on their students. If the child has learned the lesson, he or she will remember it.

Memory remains mysterious. For centuries, scientists believed that memory must be located in some specific area of the brain. In the late nineteenth century and the early twentieth century, a number of studies indicated that no matter which parts of the brain were damaged or destroyed, memory was usually left intact. Karl Lashley,[54] Hebb's teacher and colleague, was the first to speculate that perhaps memories are stored all over the brain. The areas that trigger memories are now thought to be chiefly located in the hippocampus, cerebellum, and frontal lobe. However, the memory of a single event may involve thousands, or millions, of cells in many parts of the brain.

When I first started reading about the more recent theories of memory, I was reminded of my first visit to the Museum of Holography, in New York City. There I had been fascinated to see that a holographic film could be cut into pieces, and that each piece would contain the whole picture. The little pieces were not as highly detailed as the full picture, but the basic information was all there. In other words, the picture did not break down like a jigsaw puzzle, with different bits of the picture on each piece; the big picture was encoded in every part. Fractals are a similar concept. A fractal is a rough or fragmented geometric shape that can be subdivided in parts, each of which is (at least approximately) a smaller copy of the whole. Fractals are generally self-similar (the bits look like the whole) and independent of scale (they look the same no matter how close you zoom in).

Some contemporary physicists believe that the universe behaves like a hologram. Through the ages, mystics have glimpsed something of this vision of the ultimate nature of things. William Blake's poem about seeing "a world in a grain of sand...eternity in an hour"[55] is perhaps the most familiar expression of this idea in art. Memories may be stored in our brains in a similar way, the whole in all the parts. We may be able to assemble a complete memory from any little piece of it.

However our brains file experiences and learning, we see that the musicians in the classroom seem to be quite effective at getting children to retain and retrieve information. That the children have internalized the lessons taught by the musicians shows up not just in standardized tests of reading and arithmetic, where it is harder to show cause and effect, but also from week to week and year to year in the classroom. The musicians frequently review what they have taught in the past—by asking questions—and the children, by their answers, show that they have learned and retained the information. Even after a long summer vacation, we see that retention is good.

IS GRAMMAR A GIVEN? IS MUSICAL SYNTAX INBORN?

What is going on when a third grader listens intently to a piece of music by Shostakovich? Even though the child has never heard this particular piece before, there's something familiar about it. Even though the polka form may be new to him, he recognizes that it is music, and not just foreign sounds. If one of the quintet members were to play a wrong note, the child might recognize that as well. This knowledge has not come about as a result of living for eight or nine years. Even tiny babies know music from noise, and the ability is probably innate.[56]

If language and music both seem to be built into the brains of humans, are we born with the structure of language and the structure of music already wired? The evidence is certainly intriguing.

About half of the brain is dedicated to language and the cognitive functions that use language.[57] In fact, language is the most distinguishing feature of our species. It used to be believed that babies came into the world knowing nothing. Their brains were regarded as blank slates, or tabulae rasae. John Locke, and the school of philosophy that came to be known as empiricism, believed that all knowledge and abilities were learned from experience.

In their work on universal grammar, both Noam Chomsky[58] and Steven Pinker[59] base their theories that we are hardwired for language on the observation that infants figure out the structure of language far more quickly and easily than would be expected if they had to learn and memorize the rules. Every day, a typical three-year-old correctly constructs sentences he has never heard.

All languages have certain common features, including object words (nouns), action words (verbs), and structure words (prepositions, conjunctions, and the like). Modern imaging techniques have shown that verbs and structure words are processed largely in the forward parts of the brain, and nouns are processed in the posterior associative cortex.[60] Similarly, the various elements of music are processed in many different parts of the brain: frequency is processed in the left auditory area, melody (melodic shape) in the right auditory area, rhythm in the forebrain, and much of the other work of the performing musician on the left side.

These similarities between language and music again point to the possibility that music may also be an innate language of the brain. Beth Bolton's work with babies, which we have alluded to earlier, suggests something of the kind. The infants, hearing a song, would invariably coo back one of two predictable notes, either the tonic or the dominant note of the melody. Other studies also suggest that our brains are set up for tonic and dominant sensitivity. If one note is sustained (held for several seconds), it sets up the expectation that it is either the tonic or the dominant. Again, brain imaging seems to confirm this. When just the first and fifth notes of a scale (a perfect fifth) are played, neural activity is triggered corresponding to *another* note, an octave below the first note. If the tonic and dominant notes are sounded loudly and simultaneously, this phantom sound, called a resultant tone, can actually be heard. Even though we don't hear the sound, our brains act as though the sound is occurring. Our brains fill in the blank.

The tonic and dominant idea and relationship are important organizing features of music, especially in Western music. Taken to its most developed level, the relationship of tonic and dominant is fundamental to the classical symphony, where the first theme is in the key of the tonic, and the second theme is in the dominant key.

FROM POLKA TO PROTOTYPE, FROM PAVANE TO PARADOX

What is going on when an elementary school student decides that Fauré's Pavane sounds sad and a little spooky?

Some pretty fancy thinking!

Mary Siebert, the curriculum coordinator at the Arts Based Elementary School, where the woodwind quintet is in residence, points out that children in second and third grade are at the developmental stage at which they are able to move from concrete to abstract thinking.

"When a man blows into a horn and asks, 'Is this happy or sad?'—that's a big leap," she says. Music requires this kind of movement from a sound to a state of mind, and from a state of mind to a sound.

Listening to music, with instructions about particular things to listen for, seems to help children go from sound to sense. It helps develop their ability to think abstractly.[61] This was also an implication of Dr. Bowles's finding that the curriculum improved spatial-temporal integration, as measured by the AVI test.[62]

Findings from the worlds of education and brain science might help us understand how the musicians' teaching techniques contribute to creating and reinforcing new neural networks and developing the higher cortical functions.

Barbara A. Wilson, who had been a special education teacher and a reading therapist, developed a system for teaching reading to adults and older children who had not been able to learn by standard methods of instruction. Published in 1988, that method has been used to help many children and adults who have been diagnosed with dyslexia or other reading, writing, and spelling difficulties. One of a number of reading programs based on research findings, the Wilson system succeeds in part because the instruction is *sequenced, cumulative,* and *multisensory.*[63]

Brain research indicates that new tasks and concepts are learned efficiently when instruction involves frequency, intensity, cross-training, and motivation and attention. That means the students must be exposed to the material over and over, that they must practice the new skills in a concentrated manner, that comprehension is enhanced when related skills and concepts are taught simultaneously, and that maintaining an interest in the material matters. Joaquin Fuster puts it this way (emphasis mine): "*Contiguity, repetition* and *emotional load* seem to be the most decisive strengtheners of synaptic contact in the making of a cognitive network."[64]

Musicians may not use the words *sequenced, cumulative, multisensory, contiguous, repetitive,* and *emotionally loaded,* but these adjectives add up to a pretty good description of their own musical education and training. Musicians know from experience that music is best taught in an orderly sequence, and that new information is learned most easily when it builds on what has been previously presented and learned. That's exactly the way their musical training proceeds; it is sequenced and cumulative. Musicians also know from their own training that none of the elements of a subject can be fully understood in isolation. They quite naturally combine many components and aspects of a topic in every lesson they teach. It's second nature to them to cluster and combine things that "go together," or that touch each other, and so they have a deep understanding of *contiguity.*

Repetition is of the essence to a musician. No matter how high they rise in their profession, musicians never stop practicing. Whether or not they are preparing for a particular performance, career musicians must practice every day, often for five or six hours, or risk losing much of their quality. Without regular, frequent, extensive practice, musicians lose muscle tone, necessary calluses, breath control and support, range, tone, and more. And practice is simply another word for repetition. In French, in fact, the word for practice is *répétition.*

What musicians know is that practice doesn't make perfect but that *practice makes permanent.* That's why music teachers tend to be so picky about their students' posture, how they place their fingers, when they take breaths, and so forth. It doesn't take many repetitions to lock in a particular way of doing something, and an awkward, limiting way of getting the task done can be very difficult to unlearn, or to replace with a better technique. Music teachers, and dance teachers too, often prefer to get students before they have had much or any training, because these students are less likely to have ingrained bad habits and faulty techniques.

And finally, there is "emotional load." No one needs scientific proof that we learn something more easily when the process is pleasant. For a musician, nothing is more enjoyable than playing music. Our quintet members carry into the classrooms a deep love of music. The music they play is at the very least a "pleasure technology," a time-honored way of lifting spirits and lightening tasks.

So what is going on with these third graders, lying on their stomachs, listening for images as the music washes over them, then writing down the associations and emotions suggested by the sounds? A sequenced pattern of sounds that is meaningful because the children probably came into the world with its grammar encoded in their brains. A set of instruments whose individual and commingled voices have been heard repeatedly over a period of weeks. An exercise that builds on the children's knowledge of the elements of music, starting with musical opposites. A task that requires them to connect something to something quite different, a sound to an internal state or remembered experience. A challenge to thought that is bathed in the pure pleasure of music masterfully played.

What is going on is learning.

CHAPTER 13

Young Composers

"I'm going to write a melody," Cara tells the second-grade class.

On the board, she starts on the left and draws a horizontal wavy line that looks like a range of hills. The line is continuous in some parts and broken into pieces in others. "Who'd like to tell us how to play this?" she asks.

Marcus volunteers, and comes to the front of the class and stands by Cara.

"Pick an instrument to play this melody," Cara says.

Marcus chooses the clarinet.

"Would you like Eileen to play this forte or piano?"

"Forte!"

"And *forte* means...?"

"LOUD."

"I'm writing an *f* here so Eileen will know this piece is loud. Is this piece fast or slow?"

"It's fast."

"Okay. I'm writing *fast*. Now let's see if Eileen can play it."

Eileen comes up and looks at the line on the board. She plays it through once slowly, then again at a fast and more confident tempo. Everybody applauds.

"Sometimes it helps to practice things slowly," Cara observes. "Now who would like to come up and write a melody of your own?"

All the children raise their hands. Before the half-hour lesson is over, each of them has written two or three lines on a page of his or her own. They realize that when the line goes up the sound will be higher, and when it dips it will be lower. They know that a smooth line indicates that the notes are to be played legato, and a broken line tells the musicians that that section is staccato. They indicate the tempo by writing *fast* or *slow*, and the dynamics by writing *f* or *p*. They each decide which instrument they would like to play their piece, and write the name of the instrument on the page.

As the children try their hand at composing, the quintet members look over the boys' and girls' shoulders to make sure that what they are writing will be playable. The children learn that the music has to move across the page from left to right, with no doubling back, and that because the instruments can play only one note at a time, they can't stack two or three notes on top of each other.

The kids print their names on their compositions and the quintet members gather them up, with the promise that they will perform each child's work next time.

"Do you know what people who write music are called?"

"Conductors?" one child suggests.

"Not conductors, but a word that sounds something like that."

"Composers," a little boy says confidently.

"Yes, they are composers. And that's what you are now. You're composers."

The next visit from the quintet starts with a student composition—one of the rare exceptions to the practice of beginning the lesson with a piece played by the whole quintet. The kids' compositions are stacked on a music stand at the front of the class, and Bob reads the first composer's name.

"Kershawn, will you come up here please and stand by your musician. Kershawn has written a piece for the flute."

Kershawn walks shyly up to the front of the room, with his hands in the pockets of his tracksuit, and stands next to Debi. He follows the lines on his paper intently as Debi plays. When she stops and smiles, the class applauds wildly. Kershawn looks amazed. Debi hands him his composition, to take home and keep. When he gets back to his seat, he gives the little girl next to him a high five and then goes back to scrutinizing with apparent wonder the lines he wrote on the paper.

One by one, the children come forward. They beam with pride and pleasure as the quintet members turn their pencil lines into surprising sound. Josh wrote a fast and forte piece for horn. Brianna wrote a soft piece for flute. Marcus wrote the first line of his piece for oboe and the second line for horn.

"Take a bow," Bob says. "The composers get applause, too."

We teach the Young Composers lessons at each grade level, and they seem to make a deep impression on the children. Time and again, we see children looking shy and uncertain as they go to the board to write a composition, or stand beside the musicians who are "reading" what they wrote. Over and over, we see children's eyes widen as they watch a musician transforming their marks on the page into a unique sound. In every class, we see children's chests swell and posture straighten as they realize the enthusiastic applause is for them and for what they have created.

Self-confidence can only help kids in test taking and academic achievement, and Robert Franz built esteem enhancers into the lesson plans at the earliest stages of the Bolton project. The Young Composers sessions appear to be quite effective at developing confidence and a sense of self-worth. For some children, it seems that having adults listen to them and take their ideas seriously is a new or rare experience. Some may have had little opportunity to make something, and to make something happen for all to see and hear. Some may be unaccustomed to receiving recognition and praise for their efforts.

This is certainly one of the lessons in which the playing field is leveled. None of the children have composed music before these visits from the quintet, and everybody has an equal shot. There are no obvious A, B, or C compositions. They're all different and all interesting. Everybody tries and everybody succeeds.

FROM SOUND TO SYMBOL; FROM SYMBOL TO SOUND

Important as the implicit lessons about creativity and confidence are, they are not the main goal of the composing classes. Nor is the introduction to graphs the main point, though this knowledge will serve the children well in their math courses in later grades. (Unbeknownst to them, in graphing a melody, the children are plotting pitch on the y-axis and time on the x-axis.) The primary purpose is to help children see that a sound can be represented in a visual way that then can be "read."

Not all primary grade children understand this fundamental idea. Some have been confused by learning the names of the letters of the alphabet at an early age, a practice that is discouraged in some European countries. In Montessori schools, among others, children learn the basic sounds of the language—the phonemes—before they learn letter names, and are therefore less likely to make common errors in decoding, such as pronouncing the names of the letters when trying to sound out words.

Diane McGuinness argues convincingly that widely used methods of teaching reading essentially put the cart before the horse, making the code (the alphabet) more important and real than what is being encoded (the language). To borrow her vivid example, if you think letters have sounds, try holding this page to your ear![65]

Graphing melodies gives children the opportunity to encode sound without worrying about how to spell words or punctuate properly. It's a pure and clear exercise in matching one sense (sight) with another (hearing), which is fundamental for learning to read and write. In recent years the quintet has been experimenting with having the children individually sing the melodies they have written, using no words, just singing ah-ah-ah or la-la-la. In this exercise the kids put themselves in the place of the musicians and decode what they see on the board or the page.

Most children take to this activity fearlessly and enthusiastically. They seem quite confident that they can "read" what they have written (though in fact they tend to confuse high notes with loud sounds). The musicians sometimes talk about stage fright during these exercises in which children volunteer to come up and write on the board or sing a melody they have composed. It's a safe bet that some of the children who don't volunteer to "perform" in front of the class do have some fear or nervousness. And orchestra musicians are masters at coping with performance anxiety.

MAKE BELIEVE YOU'RE BRAVE

The quintet members ask the children to think about times when they have been scared about having to do something, and to describe those times. They ask if the kids have any tricks or techniques they use to calm themselves down. Then the musicians talk about what *they* do when they get panicky, before a concert or on the stage.

We find that teachers are now very keen on this part of the curriculum. The huge emphasis on end-of-grade testing, and the high stakes involved, are stressful for everyone, faculty and students alike. Anything that will help the kids cope with the enormous pressure of taking tests is of great interest these days. I personally hope that we will stop subjecting our children to this kind of pressure, and I am supportive of the approach the Arts Based Elementary School is taking—no tests except the state-mandated tests at the end of the school year.

For performers, stage fright is a fact of life. As long as we care about doing a good job, we feel some degree of performance anxiety. For a conductor, the first rehearsal for a concert is usually an occasion of high anxiety. For many of the players, too, this period before much practice has gone into the works on the program stirs up the most jitters. Here's how we musicians handle it, and what our quintet teaches in the classrooms:

The first indication of stage fright is usually an adrenaline rush. If we identify that heightened attention as fear, it can lead to panic and paralysis. So what we do is to name that feeling "excitement." We teach ourselves to use that extra rush of energy to enhance our performance.

Another physical symptom of fear is shallow breathing. The feeling of being out of breath is as devastating to a musician as it is to an actor or public speaker. We counter it by taking slow, deep breaths. We inhale so deeply that we feel it in our lower abdomen, and we fill our stomachs with the air we breathe in.

We also tame our fear by making our muscles feel heavy. It's especially effective to imagine our faces and hands as getting heavier.

And long before we set foot on the stage, we do our homework. Being well prepared leads to confidence that we can do well, and confidence curbs anxiety. Musicians practice, practice, practice. We practice a piece over and over until we can play it almost without thinking.

FROM THE WHOLE TO THE PARTS AND BACK AGAIN

One of the ways we musicians practice is to think a piece of music through in our minds. Edwin Gordon,[66] a former teacher of mine at University High School in Iowa City, and now a professor emeritus at the University of South Carolina, coined the word *audiation* to describe this particular kind of abstract thinking. Since it requires no sensory input, audiation qualifies as a higher cortical function. Musicians who are listening to a piece of music in their heads have been shown to activate the hearing and motor parts of their brain. I find that when a piece of music comes into my mind late at night I have to "listen" to it all the way through before I can really get to sleep. In that half-asleep state, I sometimes get roused to full wakefulness by some big conducting gesture the piece demands. Along the same lines, I sometimes wonder why *Pictures at an Exhibition* is playing inside my head. Or why on earth I am tuned in to "Frère Jacques." If I search my mind, I usually discover that the name of the piece was casually mentioned to me earlier in the day. Just the name of a piece of music can be all it takes to strike up the band in my head.

This kind of internal review (audiation) is excellent for putting the parts of music in perspective, and thereby giving meaning to musical sounds. The individual notes in a score may appear to be equal, but not every moment in a piece of music is equally important. In every piece there is a peak, the moment of maximum tension, and everything else is arranged in relation to this climax. Composers intentionally build their works phrase by phrase, as a mason builds a cathedral, stone by stone. The climax in classical music most often occurs about two-thirds of the way through a piece, and it is often but not always the loudest moment. Typically, the last third of the piece is given to resolving the tension, and landing the listener in a different emotional place than he was before the piece began.

My mentor, Sergiù Celibidache, called this process the phenomenology of music,[67] referring to how perception changes as the result of the interaction between the perceiver and the perceived. Suzanne Carreker, a reading expert and the director of teacher development at the Neuhaus Education Center in Bellaire, Texas, uses the word *metacognition* in connection with the internal dialogue that good readers have with the text they are reading. She says that some children are not aware that reading is an interactive process. Teachers can help children develop metacognitive skills, in part by modeling them. Teachers and parents can show how they get personally involved with a story or article—questioning the concepts and conclusions, wondering why something has happened and what will happen next, pulling in their prior knowledge of the subject to connect ideas and bits of information.

We musicians are in the somewhat unusual position of having to constantly keep in mind the whole and the parts, the big picture and the details. Most important work—parenthood springs to mind—benefits from a person's being able to focus on both the forest and the trees, but playing music virtually demands it. As we play, we hear what we are playing and we constantly modify the next bit in light of what we have just heard. As we play, we are always aware of where the piece is going. At the beginning we already know the outcome, and the technique of audiation helps us understand the whole shape of the music.

Another word for what we are talking about, of course, is *context,* which children must grasp in order to turn words into meanings. Our musicians in the classroom communicate the idea of context with their instruments. Each note they play is part of a phrase. Several phrases in sequence make a melody. In classical music, the melody fits into a larger section of perhaps 16 or 32 bars, which may form the first theme of a movement of a big work, such as a concerto or symphony, or the second theme, the development section, the climax, or the recapitulation. Each individual note may belong to the main melody of the section, or it may belong to a harmony or countermelody.

It's important to know which, because a note played too loud is as out of place as a wrong note.

Good composers are able to weave phrases and melodies into progressions that have very few weak connections. When the links, or transitions, are weak, it's up to the performers to take care of them in a way that doesn't interrupt the flow and direction of the music. For music to feel emotionally engaging and satisfying, composers must incorporate the drama of conflict or contrast. In the absence of enough tension, listeners feel that nothing much happens. Occasionally a composer will see that a movement lacks the conflict or contrast it needs and may add it in a coda (literally, "tail") at the end of the movement. A famous example is the first movement of Beethoven's Fifth Symphony. It's likely that Beethoven realized, through audiation, that he hadn't made his point in the first movement. He added a long coda to it, and that's where the climax is.

FORMAL ATTIRE

In just about any kind of human expression, whether it's a storybook, a sitcom, or a symphony, familiarity with the underlying form is helpful to comprehension. If we go to an opera looking for the realism of a television documentary, we will miss the point. If we condemn a newspaper editorial for being opinionated and one-sided, we have misunderstood the category to which it belongs. Little children always have considerable confusion about what is real and what is imaginary, what is fact and what is fantasy. When they reach school age they become able to understand forms of expression and interpret various kinds of communications accordingly and appropriately.

Since the beginning of the Classical period, in the 1700s, art music has been composed in four basic forms, infinitely varied by the genius of individual composers:

Sonata form, also called first movement form, sets up two contrasting or conflicting ideas. Each of these ideas, or themes, is presented in the first third of the piece, called the exposition. In the middle section, the two themes interact with one another in a way that often sounds like a conflict or argument. This section is called the development, and the end of it is usually the moment of maximum tension. The last third of the form, called the recapitulation, acts as a resolution. Here the exposition is restated, with much of the conflict removed or resolved. One way to do this is to present the two themes in the same key, rather than in two different keys, as they first appeared. The two themes remain recognizable, but they have undergone a change that has reconciled them. Sound familiar? Yes, it's the same form that underlies every romantic comedy (and much other entertainment, art, and philosophy, as we will see).

A *rondo* is almost the opposite of a sonata. In this form, similar tunes are played one after another, always coming back to a refrain. Rondos usually don't generate much tension—for this reason they often take their place as the last movement of a symphony.

The *aria* form, or A-B-A form, presents a first theme, then a contrasting section, and then repeats the first theme, a little like a simplified sonata. We hear this form in many songs and in the second and third movements of large classical works. There's not much conflict in this form, just pleasing contrast. Minuets are a good example, and most dances and marches are cast in an A-B-A form.

Fugues take their name from the Latin word *fugare*, to flee, which gives us our English word *fugitive*. And that's what they sound like, one melody chasing after another. Bach's many fugues for organ are the most familiar examples of this form. The climax of a fugue occurs at the point when the key of the fleeing melodies is most removed from the home, or original, key.

FROM SOUND TO SENSE

Individual notes are part of phrases are part of melodies are part of movements are part of symphonies, just as phonemes are part of words are part of sentences are part of paragraphs are part of essays are part of books. Teachers today find that some children can decode words and read fairly fluently—without understanding anything they have read. Kids have to understand context to comprehend what they read. Just as not all notes in a score have equal importance, not all words on a page have equal importance. Kids need to be able to differentiate the main idea, the main character, and the main action from the supporting ideas, the minor characters, and the subplots.

Music is helpful in getting across the concepts of context and proportion, foreground and background, figure and field. As musicians, we are trained to hear the melody dominating the accompaniment, and to balance those elements in a pleasing manner. We bring out countermelodies, accent dissonances, and subdue sustained notes. And it is really only in *live* music, not recordings, that this lesson leaps out. Ever since the invention of the microphone, we've had the ability to manipulate sound in a way that confounds context. When you hear Frank Sinatra on the radio, his voice has the intimacy of a lover whispering in your ear, yet it can be heard throughout the house.

Before the microphone, things were clearer! A whisper was the sound of mothers and babies, and the voice of pillow talk. A soft voice was the sound of one-to-one personal communications. A louder voice was put on for business meetings and cocktail parties. And a booming voice belonged to the political candidate or the public preacher.

The intrusion of television adds to the idea that anything goes, anywhere. No wonder our kids are confused about context. We have to teach them about what clothes, words, and behavior are appropriate in various settings. If they don't know about context, they risk being teased, misunderstood, judged harshly, and, later on, excluded from educational and occupational opportunities.

When we teach youngsters about "inside" voices and "outdoor" voices, we are giving lessons in context. Whether they learn about it from parents, teachers, or musicians, children need to grasp the notion of context to be able to comprehend what they read. Without a sense of context, words on a page are baffling and ultimately boring.

When our Young Composers see the musicians playing softer and louder, when they see how one or two members of the quintet may carry the melody while the other instruments are less prominent, when they observe that in every piece of music certain things stand out and others recede, they are absorbing lessons in context. When they can sort things out this way, an experience becomes comprehensible and *meaningful.*

CHAPTER 14

Once Upon a Time

The classroom is getting ready to have new ceiling tiles installed, and until that's done it's too noisy an environment for chamber music. The musicians have set up their instruments in the big multipurpose room instead. As the third-grade class troops in, Rejine, a pretty, stylishly coiffed little girl, goes to Kendall and hands him a white box decorated with a red heart. The bassoonist opens the gift and finds a small crocheted basket.

"I made it for you," says Rejine, "because you're the best base-soon player. You're like a daddy."

Kendall is visibly touched and gives Rejine his characteristic lopsided smile. It was probably this wide, childlike grin that made Rejine think of him as a trustworthy father figure in the first place. She takes her place on the floor just as Bob walks to the front of the class.

"Do you guys ever watch TV?" Bob asks.

The yes vote is unanimous.

"What are some things you like to watch?"

The children name their favorite shows. *Rugrats. Scooby-Doo. SpongeBob SquarePants. Powerpuff Girls. Rocket Power.*

"Let's take SpongeBob," Bob says. "What do you find out when you watch that show?"

"He's a sponge."

"Right. Where does he live?

"The ocean."

"Yes. He lives in the ocean. That's the setting. Who else is in the show?"

"Mr. Crabs."

"Who else?" Bob asks.

"Sandy and Patrick."

"Yes, Mr. Crabs and Sandy and Patrick are the other characters. What is something that happened to these characters, or some kind of problem they had to figure out?"

"Don't go near the hooks or they'll put you in a tuna can and take you to the gift shop," a boy answers.

Bob tells the kids that the elements of each episode of SpongeBob are like the elements of stories in books. In all stories, there's a setting, characters, some action, and then a conclusion.

Bob reads a short picture book, *Harriet and the Garden*, by Nancy Carlson, to the class.[68] The children then identify the elements. The story's setting is a bright summer day. The characters are Harriet and her gardening neighbor, Mrs. Whozit. The problem comes when Harriet, playing ball with her friends, destroys many of the flowers in the garden, including Mrs. Whozit's prize dahlia. Harriet feels troubled and guilty about having ruined the garden, and runs away. The problem is resolved when Harriet confesses what she has done and then works with Mrs. Whozit to repair and replant the garden.

Story elements are part of the third-grade curriculum and so far the lesson has been essentially a review of what the children have learned from their classroom teacher. Now Bob takes the idea a step further.

"The parts of a story add up to a form," Bob says. "A form is a plan. Have you ever seen a dress form? That's a kind of form. A form is something you make something on. We have a piece of music by a man named Mozart, called Allegro Molto. It's written on a form called sonata form. Sonata form is like the form of a story or a play or a TV show. It has a beginning, a middle, and an end."

Debi comes forward and plays the very short opening of the piece on the flute.

"That's the beginning of the piece. It's like the setting, the way a bright summer day was the setting for *Harriet and the Garden*. What kind of feeling did you get from those first notes?" Bob asks.

To some it sounds happy, to other children sad or inconclusive. Bob says they'll have to wait to hear the whole piece to decide what the setting is like. Then he asks Debi to introduce the first theme of the piece. Debi plays a lively passage.

"What did that sound like or how did that make you feel?" Bob asks.

Many of the children stick their hands in the air, and Bob calls on them in turn: Excited...Like jumping around...Like when you haven't done something before and now you've just done it...Like when you've had way too much candy and you're hyper.

That music represents the first character, Bob tells them, and what the music sounds like tells you something about the character. In the sonata form, the first character is called the first theme.

"Now Debi is going to play the second theme, which is like the second character," Bob says. "See if you can tell me what is different about this character."

Debi plays a smoother, statelier passage.

The kids' hands shoot up again as she brings her flute away from her mouth. This music, they say, is more legato…is like a song that puts you to sleep…is like a ballet dancer slowly swirling around.

There's just time to introduce the next element of a story and of sonata form, and Debi plays the action-packed middle of the piece. To the children, this part suggests a chase, someone running from someone else.

"As we play the piece through, think about *Harriet and the Garden,*" Bob says. "See if you can hear the two different characters, and the problem that comes along, and how the problem gets resolved at the end."

Our story units are designed to help children understand and appreciate literature, and also to help them with their own writing. In these lessons, the children learn the elements of stories in three ways: from thinking about familiar television series, from hearing a story read to them, and by analogy, through listening to pieces of music that have the sonata form. Then the music is used to spark the children's own creativity. First graders are asked to listen to two contrasting pieces of music and to draw a picture inspired by their favorite of the two. In second, third, and fourth grade, children use classical music as a jumping-off point for writing their own stories.

Some children take to these imaginative tasks like ducks to water. For others the idea of thinking up characters, settings, and plots is obviously foreign. In every class, it seems, at least one or two kids will write about cartoon characters they're familiar with and often try to recall and write about a particular episode from a television series.

Bob Campbell is almost always the lead teacher for the storytelling units of the curriculum. As the father of three, he's the resident expert on both children's TV programs and kiddy lit. The son of an English teacher, Bob also enjoys literature himself and once considered following in his father's footsteps.

He says the fact that he often starts the class by talking about TV shows may confuse some kids. But in doing so, he is following the solid educational principle of moving students from the known to the unknown—and TV shows are definitely something the children know and are interested in.

From the standpoint of literacy, reading and writing are two sides of the same coin. Writing, actually forming the letters and creating texts, helps children learn to read, even at the earliest stages. Reading, the more the better, helps make good writers. At every stage, the skills of reading and writing reinforce each other.

GOLDEN OLDIES

In pointing up basic similarities between an episode of Scooby-Doo and a composition by Mozart, the musicians are touching on something deep in the human psyche. The philosophical basis of the sonata form is the resolution of conflict. First, two conflicting ideas are stated in different keys. These ideas, or themes, or characters are developed enough to show their differing characteristics. In the Romantic period, composers, released by the ideal of letting it all hang out, elaborated each theme extravagantly. In the Classical period, we see contrasting themes presented in their simplest and most distilled form.

In the middle section of the piece of music or the movement, called the development, segments of the theme are treated almost as personality traits of characters. We see these traits explored, exploited, repeated, and in various ways contrasted with the traits of the other theme. The "arguing" of the two conflicting themes creates tension and reaches a point at which the listener knows that something's got to give. The tension must be resolved in some way. There is a sense of inevitability and inexorability.

In the final section, the two themes or characters come back, but they are no longer at odds with each other. Even though each is recognizable as itself, sharp contrasts, such as different keys, are removed. The interaction of the two themes in the development section has resulted in a satisfying transformation.

This fundamental form recurs again and again in art and entertainment. Screenwriters know that the end of the second act—about two-thirds of the way through the plot—should be the lowest point for the hero and the highest point for the villain of the piece, or the moment when the lovers are farthest apart and appear unlikely ever to be reconciled. Screenwriting courses often specify to students the exact page on which that moment of maximum tension should occur. In popular entertainment, form thus often becomes formula.

For centuries, artists of all kinds have used something like the sonata form more intuitively. As an example, my collaborator, Janet, revisited her dog-eared copy of *Pride and Prejudice,* Jane Austen's great novel whose title forecasts that it will be about a conflict between two contrasting themes. The novel's two main characters, Elizabeth Bennet and Mr. Darcy, are differentiated by social class and at odds with each other because of perceived personality traits. At the end of the book, they are married. Most readers would say that the climax occurs when Elizabeth's sister Lydia elopes with the amiable scoundrel Wickham. In Janet's old Modern Library College Edition, the novel begins on page 3 and ends on page 320. Lydia's elopement is revealed on page 222.

A recent film version of *Pride and Prejudice,* with Jennifer Ehle and Colin Firth, has been shown many times on the A&E television network. Viewers can watch it with deep satisfaction time and again, just as many people read the book more than once with undiminished pleasure, despite knowing how the plot turns out. This is the same kind of deep enjoyment that listeners take in hearing a great symphony by Beethoven, Brahms, or Mahler. There is something fundamentally compelling and satisfying in the basic structure, in the resolution of a conflict that for minutes or hours has been made to seem monumental and all-important.

The placement of the climax of a novel or sonata form corresponds closely with the golden mean, a ratio known to the ancient Egyptians as the divine equation or divine proportion. Expressed in a number series by the fourteenth-century mathematician Fibonacci, the proportion, roughly two-thirds to one-third, is all around us—in architecture, in nature—and is repeated many times in our own bodies.

Sonata form and its analogs never go stale on us. No matter how often we meet this sequence of exposition, development, and recapitulation, this building of tension and its release, we continue to be emotionally engaged and satisfied by it. Familiarity with this form makes many of our experiences—including experiences of music, drama, opera, film, literature, painting, and sculpture—more coherent and meaningful. We think that for children to explicitly encounter this fundamental structure in music, as well as in stories, in the primary grades will serve them well throughout their education, throughout their lives.

ONCE MORE, WITH FEELING

It is in the story and writing lessons that the musicians deal most explicitly and directly with emotion. When they ask kindergarten and first-grade children about how stories and pieces of music make them feel, the answers are primitive and predictable. *Happy, sad, scared,* and *mad* are the words children have to describe their emotions. In the second, third, and fourth grades, the kids have many more words for talking about emotion. They may say a piece of music gives the feeling of being *excited* or *lonely*. They may recognize that Harriet feels *guilty* when she accidentally destroys Mrs. Whozit's garden, and *relieved* when she confesses what she has done.

Little children experience pleasure and pain. They react to the world with pleasure, anger, and fear. These emotions belong to the oldest part of our brains, and it's easy to see how our species needed them to survive—to avoid danger and form social attachments.

When children get to be school age, they are beginning to do something more than simply react with tears or laughter, anger or fear. They begin to be able to think about these reactions. The primitive emotion with the thought attached is what we call a feeling. When thinking comes into the picture, we can say we feel "resentful," "wistful," "ambivalent," "jealous," "nostalgic," "proud," and a whole range of emotional states. We can think about our primitive emotions.

However many *words* we have for how we feel, *music* is better at describing emotion. This ability to express emotions is the primary appeal of most kinds of music. What we see in the classroom is that musicians and music help children develop their emotional vocabulary, at the stage of development when this is first possible for them.

Although for centuries many philosophies have tried to divorce reason and passion, it doesn't work. We human beings are feeling creatures. Our emotions color what we do, and how we do just about everything. All the quintet's lessons point up the truth that people can think and feel at the same time.

The musicians understand that children learn best when they aren't feeling intimidated, anxious, inadequate, bored, or stressed. They don't give tests. They don't put kids on the spot. They don't lecture. They don't compare one child with another. They don't embarrass children or put them down. They do create an atmosphere of affection, pleasure, appreciation, and acceptance.

THE BEAUTY PART

Emotion! Imagination! Form! We seem to have come to the art part. Bringing feeling and imagination—as well as intellect—to a variety of forms of expression is what artists do. Our quintet members infuse their teaching with their artistic values.

One of the strongest messages they send is that there is rarely one right solution, or just one way to think about something. A passage of music may evoke a different image in the mind of each child in the class, and every one of them is right. A piece of music may sound sad to one child and scary to another, and they're both correct. Different members of the quintet have different ways of explaining any concept, and that's an advantage.

Another value that comes across clearly in the lessons is a respect for individuals and individuality. When a child is speaking, asking, or answering a question, or graphing a melody on the board, every member of the quintet watches and listens closely and attentively. When one member of the quintet is speaking, the others will fix their gaze on that person and join in any activity the person is leading, such as moving high and low body parts.

Even when all the musicians are not playing their instruments but are essentially doing the same thing, like clapping rhythms, they take pleasure in doing the thing in their individual ways. When the quintet was first formed, classroom teaching was a new experience for almost all the musicians, and they begged Robert Franz, the project's coordinator, to give them detailed direction on how to teach. Robert was insistent that they find their own instructional styles. He was convinced that the ability to communicate is intimately bound up with personal authenticity. He was certain that the way for them to become better teachers was to become more themselves, and to be comfortable with who they are.

That idea—that the most eloquent expression comes out of self-knowledge, self-acceptance, and sincerity—is shared among all kinds of artists, as is, I think, a delight in the specific and particular. The musicians in the class exhibit honest enthusiasm for small details and interesting little facts. They seem curious about the ways and whys of all sorts of things. In this curiosity and enthusiasm, most of them are childlike, and this may be one of the reasons the children so enjoy their visits.

Another value that comes straight from the musicians' lives as performing artists is their strong desire to communicate. They are not content to do their thing and get off the stage. They want to see that they have affected people. In the classroom this translates into their obvious belief that they have not taught until the children have learned. They will patiently explain anything in any number of ways until it is clear that the children understand.

When you don't have to focus on finding the one right answer, and when you have ample permission to be your own unique self, learning and life are more varied, interesting, exciting—and beautiful. All the artists and musicians who have had any experience with our classroom residencies are especially tuned in to the beauty part.

"I cry every time I watch the quintet in the classroom, and it's not because of how well they play," says Mary Siebert, the curriculum coordinator at the Arts Based Elementary School. Mary started piano lessons at age five, then learned to play violin and guitar before she started serious voice training at fourteen. By the age of seven, she knew she wanted to be an opera singer, and that's what she was until she suspended her touring and performing when her daughter was born.

"The most valuable thing the quintet does for the kids is not being measured and is not measurable," Mary says. She recounts this anecdote to explain what she means: A little girl transferred into the school in the middle of the semester. She'd had bad experiences at her previous schools, partly as a result of her antisocial behavior. Virtually abandoned by her parents, she had a dismal home situation. By the time she enrolled in ABES, the quintet had been visiting for some weeks and the other children in the class were accustomed to the sounds of music in their school day. The new little girl seemed to open like a flower the first time she heard the quintet. Mary, who was sitting behind her that morning, watched as she raised herself out of a slump and turned around with an expression of amazement and joy.

"She turned around and looked at me with this huge smile, like 'I can't believe this!'" Mary remembers.

The literally "uplifting" experience Mary describes is one we see often. It is most touching when noticeably troubled children at least temporarily seem to gain the ability to get outside their problems, and feel something happy and hopeful.

"If you can give beauty to these kids, what better thing could you give them?" Mary asks.

As artists, we tend to take it for granted that beauty is its own excuse for being, but it may well be that beauty is also empowering. I like the way Robert Jourdain expresses this idea of how music is transcendent: "For a few moments it makes us larger than we really are, and the world more orderly than it really is. We respond not just to the beauty of the sustained deep relations that are revealed, but also to the fact of our perceiving them. As our brains are thrown into overdrive, we feel our very existence expand and realize that we can be more than we normally are, and that the world is more than it seems."[69]

CHAPTER 15

Beyond Bolton: Theme and Variations

Thursday, November 30, 2003

Tucson Citizen

Grades Hit Higher Note:

Music, Better Learning Linked in TUSD Study

By Sheryl Kornman

Plain and simple, retired Tucson Unified School District educator Carroll Rinehart and Tucson philanthropist H. Eugene Jones love music and want to leave the world a better place.

> With their help, TUSD students are not only learning to play musical instruments, their music lessons are helping them learn.
>
> Teaching music to children enhances their neurological development and helps them learn other subjects, say proponents of the Opening Minds through the Arts project. This week, district officials released results of a new study showing that children in OMA are testing at a higher rate than children who did not take part in the program...
>
> The germ for OMA started several years ago on a trip to Winston-Salem, NC, where a similar program is in place.[70]

After our symphony musicians had been in residence for a few years, Bolton Elementary was reclassified from an "at risk" school to an "exemplary" school. Although many innovations had been incorporated under Dr. Ann Shortt's leadership, school officials believe the music residency had the largest impact in changing the educational climate.

But we had broader and higher goals than increasing the percentage of children who could pass what are essentially minimum competency tests. Back in 1994, when we were writing the grant proposals to get funding for the project, we spelled out what we hoped to achieve this way:

1. To infuse live music into the basic curriculum.

2. To teach by example the musician's creative work process of practice, revision, refinement, and then presentation.

3. To improve the young child's ability for abstract reasoning.

4. To help make learning an emotional as well as intellectual experience.

5. To develop a student's capacity for intellectual and aesthetic discernment.

6. To make classical music not only accessible but alive to students.

7. To contribute to the body of research concerning music's impact on cognition, creativity, and the development of motor skills and higher learning skills.

Not all those elements may directly or obviously correspond to how a child does on a normed multiple-choice achievement test, but as musicians, we believe they contribute to developing thoughtful, educated, intelligent people. And that's what schools are really shooting for, aren't they?

In our original grant proposals, we also stressed our intention of documenting the project so that other schools and other school districts could capitalize on the idea. Similar programs started cropping up almost as soon as the first results from Bolton were in. The quintet did mini-residencies at two private schools in Forsyth County. Then Eileen Young, the quintet's clarinetist, teamed up with a violinist for a short residency at another public school, Brunson Elementary.

Hill, the middle school mentioned earlier, had achievement and demographic profiles much like those of Bolton in 1994. At the middle school level, there were new opportunities, including a broader curriculum to which music could be related, and new challenges, including older children who had built up strong resistance to school and learning.

To date, the three-year residency at Hill has been our only experience with adolescents. Whereas the primary grade curriculum is mainly focused on teaching children to read, write, and reason, the middle school agenda has more to do with showing children different ways of looking at the world. At this age kids are trying on identities and beginning to think about what they will do and be when they grow up.

It seems that the musicians had a somewhat different kind of impact on these older children. Pat Holiday, the principal at Hill during the first two years of the musician residency, says that the sound of classical music being played in the lobby had a calming effect on the whole atmosphere of the school. This may be of particular value to kids who are at the most emotionally turbulent stage of development. The quintet's work in the classroom opened up another world to the children, she said, an unfamiliar culture, an unfamiliar type of music, a different kind of a life, and a different vision of adult employment than most had ever encountered. Another effect, she said, was that a number of children discovered a personal desire and affinity for music and went on to study music in high school.

RAISING ARIZONA

The most ambitious offspring of the Bolton project was born in Tucson, Arizona, where it continues to thrive. Here's how Carroll Rinehart, former chairman of the education committee of the Tucson Symphony, tells the story:

"Gene Jones, who was president of our symphony board, was at the American Symphony Orchestra League conference in St. Paul. On one afternoon he had no particular session he was planning to attend, and as he was walking down the hall he spotted an empty chair in an education workshop that Peter Perret was leading..."

Quintet members Lisa Ransom and Bob Campbell had joined me for that two-hour presentation to the American Symphony Orchestra League in 1998, as did Dr. Wood. It was great fun to have hundreds of symphony board and staff people, volunteers, and musicians essentially pretend to be children as we led them through soundscapes and other fundamental lessons from the Bolton curriculum. Dr. Wood then related these lessons to what is going on in the developing brain of a child. The workshop created a lot of buzz at the conference, and the tape of the session was a hot seller. One of the people who purchased a copy was Gene Jones. He took it back to Tucson and asked Carroll Rinehart to listen to it.

"I listened to that tape 10 or 15 times," Carroll said. He has been a music educator for more than half a century and has developed a national curriculum for teaching opera in schools. The last time I spoke with him, he was creating a musical television series for young children. He responded to the information about the Bolton project in much the same way I responded to that news item about the Irvine studies I had heard on National Public Radio. He felt compelled to do something about it.

Carroll called and asked me for a copy of the Bolton curriculum, and I told him we had no detailed guide or lesson plans. Then he, the board president Gene Jones, and three administrators from the school system traveled to Winston-Salem. They carefully observed the quintet in action at Bolton. Then they went home and developed OMA, Opening Minds through the Arts. The quintet members and I were invited to Tucson for a week, where we taught three diverse instrumental ensembles how to employ the methods that had been such a success in Winston-Salem.

OMA is a collaborative project between the Tucson Unified School District and the University of Arizona School of Music and Dance. Funded with a $950,000 grant from the U.S. Department of Education, OMA now operates in about 30 elementary schools. Bolton-type ensembles are used at the kindergarten level, with classroom visits twice a week for 30 weeks of the year. In the later grades the musical infusion comes from composing and performing opera and dance, playing recorders, participating in orchestra or band, and composing music.

OMA has been selected as a national model by the Arts Education Partnership in Washington, D.C., and is under study by WestEd, an independent education research firm. WestEd recently announced a preliminary report showing that children who participated in the music curriculum improved their scores in math, reading, and writing.[71] Hispanic children showed the largest overall gains. Though the Hispanic children still tested somewhat lower than white children, the gap in achievement has narrowed. School district officials credit OMA with helping children at risk for academic failure to succeed.

The Tucson schools have faced the pleasant problem of how to find enough musicians to staff all these classes, as well as after-school programs that have taken on a musical focus. They are drawing not only on musicians from the symphony, the opera, and other area orchestras, but also on college students, retired musicians, and, in the case of the after-school programs, high school students. Carroll Rinehart believes, as I do, that it is also important to get the classroom teachers to do more musical activities and not to rely entirely on music specialists.

TECHNIQUES THAT TRAVEL WELL

Closer to home, our quintet clarinetist, Dr. Eileen Young, coordinated a music residency at Parkview Elementary School in our neighbor city, High Point. The demographics of the school were similar to those of Bolton. This program used a trio of singers from the Bel Canto Company, a professional-level regional chorus. Eileen feels that singers can be even more effective than instrumentalists in teaching academics through music. One reason, she says, is that the singers can always look directly at the children while they are producing sound, whereas the woodwind players must look at their instruments, each other, and the music. After a two-year residency 79 percent of the third graders tested at or above grade level in reading, and 76.2 percent tested at or above grade level in math. The previous third-grade class, which had no exposure to the music program, had tested about 20 percentage points lower: 54.4 percent at or above grade level in reading; 57.4 percent at or above grade level in math.[72]

A recent controlled study with Head Start classes in Wisconsin also used singing as an experimental intervention to test for improvement in spatial-temporal tasks.[73] Both this study and Eileen Young's successful program in High Point open up new possibilities for music residencies because it is often more practical and less costly to use the human voice as the instrument for teaching musical concepts and lessons.

Repeated exposure to the musicians—classroom visits extending for weeks and months of the year—was fundamental to the design of the Bolton project. Because of the expense involved, it's the element that is most difficult to reproduce. Robert Franz, the first coordinator of the project, has done more than anyone to capture some of the most valuable aspects of the Bolton experience in education programs that are shorter and less intense than musician residencies.

Building bridges between classical musicians and audiences is the focus of Robert's career. As associate conductor of the Louisville Orchestra, where he is responsible for that symphony's education programs, and as conductor of the Louisville Youth Orchestra, he uses what he learned from the Bolton experience. He believes that demystifying art and artists is key, and that personal contact with the musicians accomplishes that. He's extended the "adopt a musician" concept first tried at Bolton to the Jefferson County schools. Although these children won't be having prolonged contact with orchestra ensembles, they get to know one musician quite well through interviews and other exercises, even if only in onetime visits to schools. Then, when the children go to the big annual education concert, they have a musician they know personally, a player on stage on whom to focus, and some sense of ownership.

Robert used an Italian aphorism I had mentioned to him, inadvertently switching it around, to verbalize an idea he thinks is crucial in programs that involve musicians in schools. The Italian saying can be translated, "If you desire the mother, caress the baby." What he says is, "If you want to reach the child, touch the teacher."

When contacts between musicians and kids must be brief, Robert has found it is doubly important to get the classroom teachers on board, and to take their ideas and their feedback very seriously. In Louisville, where he leads the symphony in themed education concerts, he has made it his goal to visit all 88 elementary schools specifically to meet the teachers, get their feedback on the symphony's educational programs and concerts, and ask them how the symphony can make their jobs easier. Robert also has brief meetings with the students when he visits schools, so he sees the kids first on their own territory. On the day of the concert, he stands in the lobby of the hall and greets the children as they enter. By the time they see him on the podium, then, most of them have seen him twice before, up close and personal. The friendly, informal contact makes it easier for them to listen to and learn from the music, he thinks.

For the past seven years Robert has also worked with the musicians of the National Repertory Orchestra at the Breckenridge Music Festival in Colorado. The players are at the beginning of their careers—college and graduate students and musicians in their first orchestral jobs. Each summer Robert has taught 40 to 50 of these musicians how to work with kids in the classroom. As part of the workshop, the musicians do a lesson with children, and Robert assesses them and provides individual feedback.

Through these kinds of programs, the techniques we pioneered at Bolton are spreading far and wide. The rich, interactive, friendly exchanges that go on—even when the programs are short—are a far cry from the dry "lecture-demonstrations" that have long been the norm in programs that take classical musicians into schools.

Locally, publicity about the Bolton project has made a strong case for bringing music into schools and into the education of children. In our county, the middle and high schools are seeing huge interest in the band and orchestra programs. At one middle school, for example, more than 450 children out of an enrollment of 600 participate in one or the other.

THE WHOLE NINE YARDS

Music as an art form is a pinnacle of human culture and achievement. As such, it does not need to justify its existence. It should be learned and appreciated for its own sake. But at least until every child can actually learn to play an instrument in the primary grades, I think that extended residencies by professional musicians are the most desirable way to integrate music with the basic curriculum. The residencies take advantage of everything that we know about how children learn. Lessons spaced over several weeks permit a coherent, sequenced, cumulative, multisensory learning experience. Furthermore, the affectionate relationships that develop between the musicians and the children facilitate the learning, and give the kids additional knowledge of the world.

I have often noticed that children (and also adults) get a pleasurable charge from seeing performers in different contexts. I doubt we'll ever prove it, but seeing the musicians in different roles may have some benefits for schoolchildren.

Two recent examples come to mind. Bob Campbell was shopping at a neighborhood supermarket one day last summer, when a little boy recognized him and greeted him with great excitement.

"Hi, Bob!" he said, looking around expectantly. "Where's the rest of the quintet?"

"I guess he thought we were like the Partridge family, traveling around and doing everything together," Bob comments.

More recently, I was conducting the symphony in the education concert we perform every year for all students in the upper-level elementary grades. Children from the Arts Based Elementary School were also in the audience. Without telling the orchestra players what I was doing, I asked four members of the woodwind section and one brass player to please stand. The ABES children were thrilled and called out "Debi! Cara! Eileen! Kendall! Bob!" as they recognized their teachers and friends as formally dressed performers in the big orchestra.

MUSICIAN RESIDENCIES: HERE'S HOW

What it takes to set up programs where musicians are regular, recognized, trusted, and effective teachers in the classroom will vary somewhat from school to school and city to city. Here are the fundamental requisites:

Carefully chosen professional musicians. Good musicians who are also good communicators are the first essential. Over the years, as we've had to replace members of the quintet from time to time, we've looked especially for orchestra musicians with good interpersonal skills. Quite often, young professional musicians have not had much to do with children during their long years of musical training, and it takes them a little while to remember what eight-year-olds do and like and think about. But if they are good at listening, watching, and expressing themselves in words as well as music, they can become good teachers. Special training is necessary for them to be effective immediately, and it is best provided by musicians with this type of classroom experience.[74]

In Winston-Salem, the quintet has always been made up primarily of professional musicians who play in the symphony. Graduate-level music students could also be used in residencies, and perhaps earn academic credit. Retired professional musicians would be another potential pool. Wherever they come from, it's important that they be excellent musicians. For the residencies to work, the music must be so beautiful as to command attention and create wonder.

Partnerships. Many kinds of collaboration are clearly possible and advantageous. At least one of the partners should be a not-for-profit organization that is eligible to receive grants from various funding entities. The Winston-Salem Symphony's primary partnership has been with the Winston-Salem/Forsyth County Schools. Within the school system, by far the most important partner is the school principal, whose enthusiasm and endorsement of the residency program is essential. The individual school's music teacher is another ally our musicians actively cultivate. In addition to the public school system, we've also teamed up with private schools and charter schools.

Scientific expertise. In Winston-Salem we have a partnership with the Wake Forest University School of Medicine to study and assess the program. Many universities and medical schools have research departments and specialists in the field of cognitive development. As such, their expertise can be invaluable in designing and evaluating projects. They are usually the people who are most actively exploring how children learn. In addition, they often have access to funding sources.

Money. Music residencies are not inexpensive. In Winston-Salem our first grant, from the North Carolina Arts Council, was $30,000 paid over three years, during which time we were adding one more grade each year. The current cost of the program, at one school, is $16,000 per year, reflecting an increase in musicians' compensation. This pays five musicians for 72 half-hour lessons, and provides for about an hour of planning time for each week they are at the school. Musicians are paid at the going performing rate, which varies considerably from community to community. Over the years we have had partial funding from foundations, corporations, and most recently from the Wake Forest University School of Medicine. A potential difficulty is that many grant-making organizations will fund only new and experimental projects, not ongoing programs.

Nuts and bolts. Besides time and money, ensembles in the classroom require sheet music, stands, something to write on, and appropriate space in which to play. In their lessons, they use many examples of classical, folk, and traditional music, which can be found in published sets arranged for the particular group of instruments.[75] One member of the ensemble is designated librarian and is responsible for acquiring and maintaining the music. The musicians bring their instruments, music, and portable music stands into the classroom. They generally sit on the little chairs that are in the room, in a semicircle facing the kids. Although the school faculty is often tempted to put the lessons in some more spacious environment, or to have the musicians on a stage, we consider it important that the kids see the music lessons as part of their regular school day, not separated from the other things they're learning. The musician-teachers use a chalkboard, whiteboard, or flip chart to show how words are spelled, measures filled, and stories composed, and for many other notes and diagrams.

Other considerations. We had originally thought about having both a woodwind quintet and a string quintet. When we had to trim our plans, we decided on the woodwind ensemble. That has proved a fortunate decision because there is more variety in the instruments, and I think that stimulates more interest in the children. But many types of ensembles would work well. The ensembles need time to practice together, especially in the beginning, and time to plan lessons together, also more in the beginning. Budgeting for this preparation time is essential.

Last but not least, children need to be able to see that there is no sharp division between their lessons from the musicians and the rest of the school day. The quintet members are adept at picking up cues from the classroom environment—books, bulletin board displays, and such—and incorporating them into the lessons. We also consider it important for the classroom teacher to participate, and then to reinforce concepts the musicians have taught, wherever they seem to complement what the teacher is teaching. For many teachers this integration becomes natural and seamless. I'll never forget hearing a classroom teacher say, just after the quintet had left the room, "Shall we walk to the cafeteria staccato or legato?" And then seeing the children filing out of the room in a muffled legato shuffle.

CLOSE ENCOUNTERS OF THE MUSICAL KIND

As of this writing, plans are under way for a summer reading day camp incorporating our woodwind quintet. The quintet will be there as teachers of other musicians as well as teachers of children. The camp will be open to children in grades 1 to 4. We expect that some will be kids who are behind where they should be in reading; others will be there to sharpen their skills, with the best reading tools available. The primary teachers will be reading specialists, who will teach all children by research-based methods.

I am excited about this new development, and I know it will provide us with more information about how music—thoughtfully and skillfully integrated into the basic curriculum—might help children learn to read, write, and reason. I know it will also raise even more questions about how and why music illuminates and enhances academic disciplines.

Does careful listening to music enable kids to process information faster and thus become more fluent? I suspect it does. I believe that we are helping children draw on both sides of their brain to better enable them to learn and understand. Are lessons from the Pied Piper activating the parts of the brain where frequency lives, and improving the timing that is important for reading and many mental tasks? I suspect so, but we will need scientific research to prove this. We just collected the first evidence that our music curriculum improves phonemic awareness, and I feel sure that this is a result of the active listening that is part of every lesson. As more—and more sophisticated—studies are done, science will be able to give us more explanations about what is actually going on, and changing, in the brains of children who have this kind of exposure to music. Science may one day show us that using music as a teaching tool as we have done exercises the brain in many of the same ways that actually playing music does.

GETTING CONTROL OF A COMPLEX PHENOMENON

Thus far, the controlled experiments designed to uncover a causal relationship between music and mental acuity don't give us firm ground to stand on. In a long conversation with Dr. Ellen Winner, I learned that although several studies show that playing an instrument improves spatial-temporal thinking, nobody has yet linked that to any physical change in the brain, and nobody yet knows why music has that particular effect. Dr. Winner, a developmental psychologist who is a coauthor of the REAP report mentioned earlier, is now involved in a longitudinal study, led by neuroscientists Gottfried Schlaug and Katie Overy, investigating how learning to play an instrument affects brain development.

Dr. Winner talked with me about the difficulties of designing the "perfect" study to show how music affects general learning and thinking ability. For instance, I've often thought that the fact that music is so much fun is part of what motivates the children we have worked with to become better learners. To filter out the fun factor, the control group in any truly scientific study would have to be provided with an equal amount of a different type of education that was equally fun and engaging. The degree of the children's enjoyment would have to be measured in some way, either by adult observers or by the children's own ratings of how much fun the class was.

Similarly, we've wondered to what extent having five extra adults in the classroom contributes to the good results we see. To know whether it is music or extra one-to-one instruction that makes the big difference, we would have to design a study in which the control group had five extra adults in the classroom. Finding out how brains were actually being altered would involve getting funds and parental permission for brain imaging. Finding out whether the effects of exposure to music are long-lasting would also be quite a challenge. Since children are growing and developing all along, it's difficult to interpret retention or loss of particular knowledge or skills.

It's going to be devilishly difficult to prove that music alone could produce the kind of improvement in learning skills that coincides with our musician residencies. The very fact of observing a process often changes that process, as many people know from life experiences. My coauthor, Janet, a journalist, is familiar with the alteration of people's behavior when a reporter is in the room. Perhaps you've noticed the way that even toddlers play to the camera, or act self-consciously cute when they know adults are looking at them.

Along the same lines, investigating a phenomenon may cause it to change. A story about how a blind man tried to discover the height of the waves at the shoreline vividly describes how the act of finding something out changes the situation. The man held out his arms at chest level and found no water. He then lowered his hands to be level with his waist, and his hands stayed dry. He bent over and held out his hands parallel to his knees, but he still couldn't feel the waves. Finally he dropped his hands in front of his feet, and in making contact with the water, he created waves.

As of this writing, scientific investigation of the music/mind connection is in its infancy and often contains conflicting information. This is part of the procedure of discovery—hypotheses are proposed, tested, and revised, and theories evolve to explain the data. Eventually, from these, principles and laws evolve. Among other things, we do not know whether it is more helpful to integrate music into the academic curriculum as we have done with our Bolton project, or whether music is equally effective or even more effective when it is taught as a separate subject. We also don't know exactly what the brain learns when we learn music. We don't know whether music primes the brain for other kinds of learning. These are questions for neuroscientists.

We do know that the process and curriculum we have described here work, even though we don't know exactly which elements in this complex intervention make the big difference. For many schools, teachers, and children, a good model will be good enough. For many who urgently need to learn to read and succeed, empirical evidence will do until scientific proof comes along. I look forward to being personally involved in designing some of the research projects that will shed more light on this fascinating subject. Meanwhile, let the band play on!

As I watch our woodwind quintet in the classroom, I sometimes think of that haunting series of notes that immediately signals *Close Encounters of the Third Kind*. In that movie, the notes—two-three-one-one-five—are beamed to the universe. With the pitch sharply rising at the end of the phrase, the notes form a kind of question. And yes, the message is received. And yes, the question finds an answer.

That is what I think the woodwind quintet is doing. Our musicians are playing to a fundamental language of the brain. They are evoking a muse that already lives in every child's head.

CHAPTER 16

Coda: Play It Again

The quintet members file into the third-grade classroom and play a lively number called Gallop.

Michael raises his hand.

"That was very staccato," he says knowledgeably.

Over three years, his class has come to know the quintet, and the affection between the musicians and the children is palpable. Summer vacation is just a couple of weeks away, and as the school year ends, the class wants to thank the quintet in a special way. The children have made a large bound book of their artwork and they've prepared a song to sing.

One child goes to the CD player at the back of the room and turns on the instrumental soundtrack of the song. Another child brings the book up to the front of the class, ready to turn the pages for the musicians to look at during the song.

Softly and from memory, the whole class sings:

Come run the hidden pine trails of the forest
Come taste the sunsweet berries of the Earth
Come roll in all the riches all around you
And for once, never wonder what they're worth...

We need to sing with all the voices of the mountains
We need to paint with all the colors of the wind
You can own the Earth and still
All you'll own is Earth until
You can paint with all the colors of the wind.[76]

"Beautiful!" Bob exclaims.

"Thank you," says Eileen.

"Thank you all so much," says Lisa.

The quintet applauds enthusiastically. The children beam with pride and pleasure.

Changing the complexion of a piece of music—from sorrowful to optimistic, from weary to jubilant—is easier than one might think. Changing the mode, and with it the mood, is a matter of tinkering with only one note of the scale. Minor adjustments can lead to major differences.

And so it has been with our music residencies in schools. In the best of times, the musicians have been in each classroom for only a total of nine hours of the school year. Yet that seems to have been enough to make a positive difference.

I think that improved listening is the largest single factor contributing to the higher test scores we've seen following the music programs. But I, and others who have been closely associated with the program, would point to additional factors that probably affect the outcome, from the benefits of having five extra teachers in the room to explain things and manage the class, to the seriously playful approach of the musicians, which seems close to the way children discover and learn things on their own.

Without a doubt these itinerant musicians have freedoms and advantages not shared with teachers who are in the classroom day after day, month after month. Our quintet members have high respect for teachers, who have so much that they are accountable for, and so many restrictions on what and how they can teach. The musicians are not under pressure to "teach to the test"—and they don't. Instead, they feel confident that if they can get the children excited about learning, good test scores will be a by-product of that enthusiasm. As Bob Campbell says, "We aren't so much teaching them *things*. We're teaching them how to learn."

Being a little bit outside the system, the musicians don't get involved with the labels that, while they may help children get appropriate educational services, can become excuses for not succeeding. They don't know whether any particular child has a diagnosis of ADHD, comes from a dysfunctional family, or has a low IQ. They assume that all the children will be able to learn the fairly subtle concepts they teach. If a child doesn't understand, they take responsibility themselves, rather than blaming the child for not having learned. They have high expectations of the children, and by and large those expectations are met.

So I would say that the quintet members exhibit the characteristics of good teachers, in general, and good team teachers. They also bring to the classroom the liveliness, openness, and playfulness that might be expected from artists in any art form. But much of their value is that they are musicians, and their teaching medium is music. Not only are the musicians good teachers, they draw heavily on the model of their own musical training, the step-by-step, review-and-repeat, draft-and-revise approach that works so well in helping children grasp and remember both content and process. Not only is music an excellent instructor in sound, form, timing, symbol, and the matching of one sense to another, the live presence is virtually irresistible. Kids can't take their eyes, or their ears, off the quintet when it is playing. They are a captive audience, ready to learn.

Seeing the musicians in the classrooms, and the joyous, affectionate reception they receive, makes me painfully aware of the musical deprivation that schoolchildren have experienced over the last 20 or more years. My friend Carroll Rinehart, a leader of the exciting OMA program in Tucson, talks about the energy that music creates in schools. In some classrooms in Tucson, he tells me, when things start to feel lethargic and bogged down, the teacher and students sing until the energy is revived. I can't count the number of times I've seen an adult or a child walk by a classroom where the quintet is playing and spontaneously break into a smile, then continue down the hall with a springier step and a lightened mood.

The sound of music would make our schools more humane, more orderly, more vibrant, and more conducive to learning. It's time we stopped allowing music to be dismissed as dispensable. It's time to give all our elementary school children meaningful expression of the language whose structure is part of their brains, and whose evocation could allow them to better use their brains.

This generation of children needs to become aware that music is one of the highest achievements of culture, and that it belongs to them! They need to learn early that great music is not above and beyond them but that it can be their companion and friend for life.

Musicians who have been intimately involved with our residencies may be forgiven for believing that the knowledge and love of music is the best reason for making music in schools. Robert Franz says, "I was wonderfully shocked when I heard the first set of test results from Bolton, but that meant nothing compared to the fact that those kids could listen intelligently to Samuel Barber's *Summer Music*. It's a complex, 18-minute piece of music for wind quintet, and we played it as part of an hour-long concert, and they were totally absorbed by it. Because of their experience with the quintet, they knew, and know, how to have a quality experience at a classical music concert."

Mary Siebert, the curriculum coordinator at the Arts Based Elementary School, sees the enduring value of having classical ensembles in the classroom in much the same way. "I don't like to see the arts separated from the rest of school or the rest of life," she says. "By having the quintet here, kids become able to embrace music as a tool of their own. The music empowers them instead of distancing them. I love the intimate contact they get with these particular instruments—what a bassoon is, what a reed is, what the sound of each instrument is. All their lives those little details will be part of their psyches."

A MODEST PROPOSAL

All schools, right now, could begin to reap more of the benefits of music, without major policy changes or major infusions of money. Some music-related activities that offer academic benefits could be led by classroom teachers. Other lessons could be presented by music teachers or resource teachers, who could leverage their valuable time to make a more specific connection between music and other parts of the curriculum.

SUGGESTIONS FOR CLASSROOM TEACHERS

Here are ideas that classroom teachers by themselves, or with parents or other volunteers, could easily implement:

- Expose children to the sounds of all the instruments in the orchestra to build phonemic awareness. Nothing beats live instruments, but recordings can substitute if the real instruments aren't available. Musical works that can help children learn to distinguish the sounds of instruments include Prokofiev's *Peter and the Wolf,* Benjamin Britten's *Young Person's Guide to the Orchestra,* and Russell Peck's *Thrill of the Orchestra.*

- Sing, and have children memorize the words of songs and poems.

- Clap the syllables of poems and songs, or use claves (two sticks to beat together) or percussion instruments.

- Tell children about the lives of composers, so that they can be inspired by their greatness and moved by their humanity. (I'll never forget learning in school that Haydn had cut the pigtail off the wig of the boy sitting in front of him in *his* schoolroom.)

- Make field trips to orchestra concerts and rehearsals, and performances of all kinds of music, from folk to opera.

- Bring musicians into the classroom so that children can see them as people and understand something about their role in society. Prepare the children to interview them about their instruments, their training, and their careers.

- Learn and sing songs whose words help children manage their emotions and internalize positive values. Musicals and Disney movies are good sources of such music. Some examples: "I Whistle a Happy Tune," "When You Wish Upon a Star," "Colors of the Wind," "Whistle While You Work," "Getting to Know You."

SUGGESTIONS FOR MUSIC TEACHERS

I would like to see half-hour music lessons at least three times a week for all students. Trained music teachers might employ these tactics, if they are not already using them:

- Teach the names of the musical notes, always starting the do-re-mi scale with the same note, C (fixed do).

- 𝄞 Teach all children to play musical keyboards.

- 𝄞 Get children to think about music, by means of such exercises as asking them to predict whether the next note in a piece will be higher or lower, or what it is in the music that causes them to feel happy, spooky, sad, or excited.

ADVANTAGES OF MUSICIAN RESIDENCIES

Residencies of several weeks or months, by instrumental or vocal ensembles, can produce the most dramatic results in achievement. They can make the most difference in the school experience and success of children who have difficulty in learning and liking reading, writing, and arithmetic. As we have seen, musicians are models of discipline, teamwork, effort, and excellence. As we have seen, they are always teaching higher thinking skills along with practical information and "world knowledge." As we have seen, they more or less intuitively teach in the sequenced, cumulative, multisensory way whose success is supported by educational research and that is advocated by experts. Getting the best payoff from musicians in the schools requires exposure over some weeks, not the one-shot approach that is more typical.

As parents we intuitively teach our children through the arts—with nursery rhymes, finger paints, fairy tales, crayons, modeling clay, lullabies, puppets, xylophones, whistles, drums, dancing around, dressing up, and make believe. Human beings have never devised a better way of educating little children about the world around them and the world within them. Why would we want to abandon this engaging, enlivening, effective approach at the schoolhouse door?

We don't stop needing art and what we can learn from art at age 6—or even at age 60. More than anything, art is what humanizes us. Without it we become "hollow men," mechanical and devoid of spirit.

Of course music belongs in our classrooms. A symphony or sonata or fugue is not unlike the life of a child, an intricate interweaving of color, shape, intensity, and even simplicity, a wondrous unfolding of something uniquely beautiful. In a place and time where too much is categorized, specialized, and analyzed, music brings a salutary experience of wholeness. The richness of music is a mirror of our minds and an echo of our souls.

Do Try This at Home

Tips for Parents

You don't have to be a trained musician and you don't have to own a musical instrument to give your children the benefits of careful listening. Use these activities to attune your children to some of the subtleties of sound.

- Ask your child to close her eyes and listen carefully to all the little noises and sounds around her. After 60 seconds, ask her to open her eyes and name all the sounds she heard. This can easily be done with two or more children, and can be done at home, in the car, or in any environment.

- Ask your child, or children, to think of a particular setting—for example, an airport, a forest, a busy intersection, a farmyard, or a city street during a storm. Let him use utensils and objects from the kitchen, such as aluminum foil, containers filled with rice, or rotary eggbeaters, to create the sounds of that place.

- Play a recording or choose a music station on the radio. With your child, move to the music, moving high, middle, and low body parts as the music goes higher and lower. Try to move in the same way as the music—big movements for loud sounds, slight movements for soft sounds, gliding motions for smooth sounds, and choppy gestures for bumpy sounds.

- In the house, or anywhere you happen to be, ask your child to look and listen and find opposites. Examples: The ceiling is up, the floor is down; the trees are outside, the chairs are inside; the milk is cold, the soup is hot; the baby was crying, now she is laughing; the truck sounds loud going by, the bicycle sounds soft. Explore with your child what kinds of things don't have opposites.

- Clap a simple rhythm for your child and ask her to clap it back to you. Continue clapping out rhythms, making them more complex and faster. Ask your child to clap rhythms for you to copy back to her.

- Clap the rhythm (the words) of a familiar song to your child and ask him to guess which song it is. Repeat until your child guesses the song. Then take turns.

GLOSSARY

Allegro The Italian word for "fast."

Amplitude The height of a wave. In acoustics, it is the size of the vibration (as opposed to its frequency). The greater the amplitude, the louder the sound. See also **frequency**.

Articulation Musical term that refers to how notes relate to each other: very detached (*staccato*), smooth (*legato*), or even run-on (*slurred*).

Audiation Takes place when we hear and comprehend music when the sound is no longer actually present; to hear it in one's head.

Axon A nerve cell fiber carrying an impulse toward another nerve or a muscle.

Baroque period In music history, the period beginning with Corelli and ending with Bach, approximately 1650 to 1750. Baroque music is characterized by the use of counterpoint and is often exuberant.

Beat A usually inaudible but regular and predictable pulse that forms the rhythmic underpinning of music. In jazz, the beat is explicit and audible. In symphonic music, the orchestra conductor sketches the beat with his arms.

Broca's area A region of the brain named after the French surgeon Paul Broca (1824–1880). Located in the inferior frontal gyrus of the left frontal lobe of the cortex, it is involved in speech production and in the processing of syntax.

Cadence A moment in music usually between two phrases, in which a chord progression indicates a pause of sorts. The Italian for "cadence" is *cadenza,* which is typified by a flowery or elaborate solo passage, usually near the end of a movement.

Classical period In music history, the period following the Baroque. It begins with Haydn and ends with Beethoven, approximately 1750 to 1825. Period in which the sonata, the symphony, and the string quartet were invented.

Coda The Italian word for "tail"; musically, the end of a movement or piece. Like the tail of a squirrel or cat, it is an appendage that balances and adds grace.

Corpus callosum Latin for "firm body." It is a large band of nerve fibers that link the two hemispheres of the cerebral cortex.

Cortex Latin for "bark" or "shell." It is the outermost part of the brain, the gray matter. Also called cerebral cortex and neocortex. It comprises six layers of many kinds of neurons and is just slightly over a sixteenth of an inch thick (2 mm). See also **white matter**.

Cortical column A columnlike structure of brain cells that function as a unit.

Dominant In music, the harmonic area or a chord relating to the fifth note of the scale. The dominant, usually notated by the Roman numeral V, plays a large role in the development of longer forms of music.

Dynamics In music, the relative loudness of any note or passage. *Piano* is Italian for "soft" and is abbreviated *p*. *Forte* ("loud") is abbreviated *f*.

Dyslexia Literally, difficult reading. It is a severe difficulty in learning to read and is generally recognized as a different way of processing the written word. It has no relationship to intelligence.

Frequency In music and acoustics, it is the number of vibrations per second, which determines the pitch of the sound. Higher frequencies correspond to higher pitches. Frequency is measured in units called Hertz (for Heinrich Rudolf Hertz, 1857–1894, a German physicist). The frequency of A is written A = 440 Hz.

Frontal cortex, or frontal lobe The forward part of the cortex, behind the forehead. Planning, strategies, and coordination are processed here.

Harmonics Secondary pitches produced by a source (string, air column, etc.) as it vibrates in its entire length (fundamental tone) and in fractions. They are softer than the fundamental pitch and vary from instrument to instrument. See also **timbre**.

Higher cortical functions (HCF) Commonly refers to brain processes (thinking) that do not involve direct sensory or motor input. Abstract thinking, solving problems in one's head, and making plans and strategies are examples of HCFs.

Kindermusik A type of music class in which educators lead a group of parents and their children through activities, using music and movement. Parents learn more about their child's unique developmental process, and the shared learning experience creates a unique bond as the child associates learning with fun, musical play.

Magnetic resonance imaging (MRI) An imaging procedure that allows scientists to "see" soft tissue in the body by observing magnetic changes in the body when it is placed in a strong magnetic field. Functional magnetic resonance imaging (**fMRI**) shows where increased blood flow is occurring.

Major mode Western musical scale characterized by two whole steps, a half step, three more whole steps, and a final half step. An example would be the white keys of the keyboard from C to C. Often perceived as "happy."

Minor mode One of several Western musical scales characterized by one whole step, a half step, and two or more whole steps. An example would be the white keys of the keyboard from A to A. Often perceived as "sad."

Motor cortex Those parts of the brain that control the voluntary muscles of the body. The motor cortex runs in a band from (approximately) the front of one ear across the top of the head to the other ear. The right side controls motor function on the left side of the body, and vice versa.

Myelination The process by which myelin, a fatty sheath, is laid down on a nerve fiber. Myelinated fibers transmit impulses many times faster than unmyelinated fibers.

Neocortex See **cortex**.

Neuron A nerve cell. There are many kinds of neurons, each having different properties and functions.

Octave The eight-step interval between any two tones whose frequencies have the ratio 1:2. Notes an octave apart sound almost the same, yet higher or lower than each other.

Overtone series A series of notes called overtones, related to the fundamental note by simple ratios.

Overtones Soft, barely audible notes belonging to the overtone series; they give the fundamental note the characteristics that define it—its timbre. Differing amounts of overtones cause us to hear the difference between vowels, and between different musical instruments.

Partial Another name for an overtone, deriving its name from the fact that it is produced from a vibration of only part of the source of the sound.

Phonemes The smallest part of a spoken word. Different languages have different numbers of these basic sounds. For example, the word *cat* is made up of three distinct phonemes: /k/, /a/, /t/. American English uses about 44 phonemes; some African languages use more than 200.

Pitch Used to describe a fundamental unwavering frequency. It generally includes the overtones produced by an instrument or a voice. The oboe and the violin play the same pitch when they tune to A = 440 Hz, even though they have different timbres.

Positron-emission tomography (PET) scan An imaging procedure that allows scientists to detect areas of increased blood flow as indicated by localized release of measurable energy. Unstable positrons are injected into the body in a sugar solution.

Register A range of pitches produced by an instrument or a voice. A register is usually referred to as "high," "middle," or "low."

Rhythm A primary element of music, characterized by longer and shorter time intervals. Rhythm can be regular or irregular and is often defined as "measured time."

Romantic period In music history, the period following the Classical period. It begins with the later works of Beethoven and ends with the Modern period, approximately 1825 to 1910. The Romantic period saw the invention of the tone poem and the creation of large-scale ballets and operas. It is characterized by a tendency toward personal expression and colorful descriptiveness in orchestration, and great virtuosity in solo writing.

Solfège A singing method popular in Europe that consists of singing in rhythm, using the following note names: do = C, re = D, mi = E, fa = F, sol = G, la = A, ti or si = B.

Sonata form A three-part music form characterized by the presentation of contrasting themes or melodies in the first section (exposition), a working-out section (development) in the middle that leads to a climax, and a repeat of the first section with the contrasts de-emphasized (recapitulation). The sonata movement often ends in a coda.

Spatial-temporal reasoning Reasoning that includes space or shape perception and sequence. In adults it often is tested with rotation tasks or paper cutting and folding tests. It is very important in abstract thinking, the kind of thinking involved in, for example, the mental planning of chess moves.

Synapse Junction where an impulse is transmitted ("fired") from one neuron to another.

Tempo The speed of a piece of music. Traditionally, Italian and German terms are used in classical music to describe these speeds. A march has a faster tempo than a ballad.

Temporal lobe A large part of the brain, stretching from about the joint of the jaw to behind the ear, on each side of the head. Many auditory processes take place in the temporal lobes.

Timbre The tone of a pitch. It can be likened to color and is what typifies the tone of an instrument. The timbre of the oboe resembles that of a muted trumpet, but not that of the flute. Timbre results from the arrangement and distribution of **overtones** generated by an instrument or singer.

Tonic In music, the harmonic area relating to the first note of the scale. The tonic is usually notated by the Roman numeral I.

Tonotopic Organized so that specific areas or nerves are related to specific frequencies.

Wernicke's area Area in the left **temporal lobe** named after the German neurologist Karl Wernicke (1848–1905). It is involved in language comprehension.

White Matter The areas below the cortex, consisting mostly of myelinated fibers. Whereas at birth the human infant's brain weighs less than half its eventual adult weight, it possesses far more neurons than the adult brain. What is still unformed is the vast network of connections that crisscross the brain in every direction. These networks develop from infancy, primarily through interaction with the environment—that is, through sensory and motor input.

NOTES

1. J. R. Skoyles, "The Singing Origin Theory of Speech" (lecture, Paris: "3rd Conference on the Evolution of Language," April 3–6, 2000). Also, M. Vaneechoutte and J. R. Skoyles, "The Memetic Origin of Language: Modern Humans as Musical Primates," *Journal of Memetics—Evolutionary Models of Information Transmission*, Transmission 2 (1998), http://JoM-emit.cfpm.org (accessed May 10, 2004).

2. Isabelle Peretz, "The Nature of Music," Advocacy Article, published online 2003 at http://www.fas.umontreal.ca/psy/GRPLABS/lnmcg/website/downloads/credoLNMCA.pdf (accessed May 6, 2004).

3. Beth Bolton, *Hearing the Signal Through the Noise: Evidence of a Musical Language in Infants* (lecture and video, Pittsburgh, PA: "Harmonic Development Conference: Music's Impact to Age Three," hosted by the Pittsburgh Symphony, March 24–25, 2000). It should be noted that humans hear in utero from about the fourth month of gestation, and the effects Bolton describes could reflect cumulative listening experiences with music.

4. Jenny R. Saffran, "Mechanisms of Musical Memory in Infancy," in Peretz and Zatorre, eds., *The Cognitive Neuroscience of Music* (Oxford: Oxford University Press, 2003): 35–39.

5. Sandra E. Trehub, "Musical Predispositions in Infancy," in *The Cognitive Neuroscience of Music* (see note 4), 13.

6. Jenny R. Saffran, "Mechanisms of Musical Memory in Infancy," in *The Cognitive Neuroscience of Music* (see note 4), 39.

7. S. N. Malloch, "Mothers and Infants and Communicative Musicality, *Musicae Scientiae* Special Issue (1999–2000): 29–57.

8. D. Falk, "Prelinguistic Evolution in Early Hominids: Whence Motherese?" *Behavioral and Brain Sciences* (2004): In press.

9. F. H. Rauscher, G. L. Shaw, and K. N. Ky, "Music and Spatial Task Performance," *Nature* 365 (1993): 611.

10. Plato, *The Republic,* trans. F. M. Cornford (Oxford: Oxford University Press, 1945), 86.

11. An expanded version of the pilot study may be found in F. H. Rauscher, G. L. Shaw, L. J. Levine, E. L. Wright, W. R. Dennis, and R. L. Newcomb, "Music Training Causes Long-Term Enhancement of Preschool Children's Spatial-Temporal Reasoning," *Neurological Research* 19 (February 1997), 1–8.

Studies of students in the fourth to sixth grades did not demonstrate improved long-term quantitative or verbal cognitive abilities. Eugenia Costa-Giomi, "The Effects of Three Years of Piano Instruction on Children's Cognitive Development," *Journal of Research in Music Education* 47 (Fall 1999).

12. William Congreve, *The Mourning Bride,* act I, scene 1, lines 1–2.

13. The school counselors, using the Otis-Lennon School Ability Test, tested the second and fifth graders at the beginning of the school year. The composite or average of these scores for the Bolton students in 1994 was 92; 100 was the average for the entire school district. The school system no longer tests students' IQs unless learning or behavioral problems warrant it.

14. The End of Grade (EOG) tests, administered statewide starting in the third grade, evaluate the students' level of reading and arithmetic as compared to grade level. In fourth grade, writing is added to the testing. Since 2003, the federal No Child Left Behind testing has been added to the EOG tests as a means of ensuring that minority children's educations are not being neglected.

In 1996, the Bolton third graders' tests showed only 36.5 percent at grade level or above in reading, and 38.1 percent at grade level or above in math. In 1997, after three years of the music program described in this book, the EOG scores in grade 3 revealed that 85.7 percent of students performed at grade level or above in reading, 89.3 percent in math.

15. The Clarion Wind Quintet had come to Winston-Salem in 1965 as the quintet in residence and wind faculty of the fledgling North Carolina School of the Arts. The Winston-Salem Symphony was the beneficiary of a Rockefeller Grant that allowed it to engage the Clarion Wind Quintet as its principal wind players.

16. In this case, the researchers were focusing on small parcels of brain cells known as cortical columns, which may contain hundreds of cells having various functions and relationships to other cells. The cells in the column function as a unit. Described in Gordon L. Shaw, *Keeping Mozart in Mind* (San Diego: Academic Press, 2000), xiv–xv.

17. F. H. Rauscher, G. L. Shaw, and K. N. Ky, "Music and Spatial Task Performance," *Nature* (see note 9), 611. Elaborated in F. H. Rauscher, G. L. Shaw, and K. N. Ky, "Listening to Mozart Enhances Spatial-Temporal Reasoning: Towards a Neurophysiological Basis," *Neuroscience Letters* 185 (1995): 44–47.

18. G. Schlaug, L. Jaenke, Y. Huang, J. Staiger, and H. Steinmetz, "Increased Corpus Callosum Size in Musicians," *Neuropsychologia* 33 (1995): 1047–1055. And G. Schlaug, L. Jaenke, Y. Huang, J. Staiger, and H. Steinmetz, "In Vivo Evidence of Structural Brain Asymmetry in Musicians. *Science* 267 (1995): 699–701.

19. C. Gaser and G. Schlaug, "Brain Structures Differ between Musicians and Non-Musicians," *The Journal of Neuroscience* 23 (October 8, 2003): 9240–9245.

20. Howard Gardner, *Frames of Mind: the Theory of Multiple Intelligences* (New York: Basic Books, 1985).

21. E. Winner and L. Hetland, "REAP: Executive Summary," published online 2003 at www.pz.harvard.edu/Research/Reap/REAPExecSum.htm (accessed May 6, 2004).

22. Reynolda House, the former country estate of R. J. Reynolds and his family, is now a museum of American art.

23. There are numerous scientific papers on this subject—for example, D. Schwender, H. Kunze-Kronawitter, P. Dietrich, S. Klasing, H. Forst, and C. Madler, "Conscious Awareness During General Anaesthesia: Patients' Perceptions, Emotions, Cognition and Reactions," *British Journal of Anaesthesiology* 80 (1998): 133–139.

24. N. Herschkowitz and E. C. Herschkowitz, *A Good Start in Life: Understanding Your Child's Brain and Behavior* (Washington, DC: Dana Press, 2002), 101.

25. John J. Ratey, *A User's Guide to the Brain* (New York: Pantheon Books, 2001), 97.

26. Isabelle Peretz, "Brain Specialization for Music: New Evidence from Congenital Amusia," in Peretz and Zatorre, eds., *The Cognitive Neuroscience of Music* (see note 4), 192–203.

27. Joaquin M. Fuster, *Cortex and Mind: Unifying Cognition* (Oxford: Oxford University Press, 2003), 11–16.

28. John J. Ratey, *A User's Guide to the Brain* (see note 25), 35.

29. Joaquin M. Fuster, *Cortex and Mind* (see note 27), 36–53.

30. Ibid., 32.

31. Lynnell Hancock, "Why Do Schools Flunk Biology?" *Newsweek*, February 19, 1996, 58.

32. Ione Fine et al., "Long-Term Deprivation Affects Visual Perception and Cortex," *Nature Neuroscience* 9 (2003): 915–916.

33. W. Dixon Ward, "Absolute Pitch," in Diana Deutsch, ed., *The Psychology of Music* (New York: Academic Press, 2nd edition, 1999): 265–298.

34. Robert Jourdain, *Music, the Brain and Ecstasy: How Music Captures Our Imagination* (New York: William Morrow, 1997), 18–41.

35. Jack Prelutsky, illustrated by Marilyn Hafner, "The Turkey Shot Out of the Oven," in *It's Thanksgiving* (New York: Greenwillow, 1982).

36. S. M. Rao, A. R. Mayer, and D. L. Harrington, "The Evolution of Brain Activation During Temporal Processing," *Nature Neuroscience* 4 (March 2001): 317–323.

37. Maryanne Wolf, ed., *Dyslexia, Fluency and the Brain* (Timonium, MD: York Press, 2001).

38. K. Overy, R. Nicolson, A. Fawcett, and E. Clarke, "Dyslexia and Music: Measuring Musical Timing Skills," *Dyslexia* 9 (2003): 18–36.

39. E. L. Grigorenko, F. B. Wood, M. S. Meyer, L. A. Hart, W. C. Speed, A. Shuster et al., "Susceptibility Loci for Distinct Components of Developmental Dyslexia on Chromosomes 6 and 15," *American Journal of Medical Genetics* (Neuropsychiatric Genetics) 60 (January 1997): 27–39.

40. B. F. Pennington, J. W. Gilger, D. Pauls et al., "Evidence for Major Gene Transmission of Developmental Dyslexia," *Journal of the American Medical Association* 266 (September 18, 1991): 1527–1534.

41. "The Auditory-Visual Integration test is a computerized test used to measure spatial temporal tasks. The AVI measures visual, spatial, visual temporal, and auditory temporal modalities and cross modal stimulation in counter-balanced paradigms." S. Bowles, "Tune Up the Mind: The Effect of Orchestrating Music as a Reading Intervention" (Ed.D. diss., Indiana University of Pennsylvania, 2003), 87.

42. Ibid., 145.

43. National Assessment of Educational Progress (NAEP) report, http://nces.ed.gov/nationsreportcard/reading/results2003 and http://www.ed.gov/programs/readingfirst/faq.html (accessed May 6, 2004).

44. Discussed at length in Jane M. Healy, *Endangered Minds: Why Children Don't Think and What We Can Do About It* (New York: Simon and Schuster, 1990).

45. Diane McGuinness, *Why Our Children Can't Read* (New York: The Free Press, 1997), 10.

46. "Suicidal behavior is among the most serious consequences of depression. In a survey of high school counselors, a higher-than-expected rate of students with learning disabilities was found among youths with suicidal behavior (Hayes and Sloat, 1988). Conversely, cross-sectional analyses from the National Longitudinal Study of Adolescent Health have indicated that adolescents with learning disabilities have approximately twice the risk of attempted suicide in the last year relative to adolescents without such learning problems (Svetaz et al., 2000). In a clinic-referred sample, but not in a sample of co-twins, Boetsch et al. (1996) found that youths with reading disabilities had more suicidal ideation than youths without reading problems. Combined rates of suicidal ideation and attempts in one longitudinal study were more common among adolescents with poor reading than among adolescents with typical reading; these differences persisted after controlling for depression, and were related to eventual school drop-out (Goldston et al., 2002; Arnold et al., 2003)." David Goldston et al., "A Review of the Psychosocial Correlates of Reading Disabilities: Methodological and Conceptual Issues." Under editorial review, 2004.

47. Rudolph Flesch, *Why Johnny Can't Read* (New York: Harper Collins, 1955).

48. Steven Pinker, Foreword in Diane McGuinness, *Why Our Children Can't Read* (see note 45), ix.

49. Recent evidence indicates that many Chinese pictographs are processed phonologically. Li Hai Tan and C. Perfetti, "Chinese Word Reading," *Journal of Experimental Psychology: Learning, Memory and Cognition* 25 (1999): 382–393.

50. Sally Shaywitz, "Dyslexia," *Scientific American* 275 (November 1996): 98–104.

51. Joaquin M. Fuster, *Cortex and Mind; Unifying Cognition* (see note 27), 45.

52. Ibid, 57.

53. Joseph LeDoux, *Synaptic Self; How Our Brains Become Who We Are* (New York: Viking, 2002), 79–81.

54. Karl S. Lashley (1890–1958) was born and educated in the United States. He formulated the theory of cortical specialization for sensory and motor functions, challenging the current concept of cortical localization. The controversy between the modular and holistic views of brain functions was brought into focus by Lashley. He was Donald Hebb's teacher.

55. William Blake, "Auguries of Innocence," line 1.

56. Sandra E. Trehub, "Musical Predispositions in Infancy: An Update," pp. 3–20, and J. R. Saffran, "Mechanism of Musical Memory in Infancy," pp. 32–41, both in Isabelle Peretz and Robert Zatorre, eds., *The Cognitive Neuroscience of Music* (see note 4).

57. Joaquin M. Fuster, *Cortex and Mind* (see note 27), 180.

58. Noam Chomsky, *Language and Mind* (New York: Harcourt Brace Jovanovich, 1968).

59. Steven Pinker, *The Language Instinct: How the Mind Creates Language* (New York: Harper Perennial, 1994).

60. Joaquin M. Fuster, *Cortex and Mind* (see note 27), 193–195.

61. Gordon L. Shaw, *Keeping Mozart in Mind* (see note 16), 19–20.

62. S. Bowles, *Tune Up the Mind* (see note 41), 146.

63. Barbara Wilson, "Wilson Reading Method: Controlled, Decodable, and Enriched Text for Students with Primary Decoding Deficits" (lecture, Winston-Salem, NC: June Lyday Orton Memorial Symposium, Wake Forest University, 2003).

64. Joaquin M. Fuster, *Cortex and Mind* (see note 27), 82.

65. Diane McGuinness, *Why Our Children Can't Read* (see note 45), 74.

66. Edwin E. Gordon, noted teacher, lecturer, author, and researcher in music education and the psychology of music, is currently Distinguished Professor in Residence in the University of South Carolina School of Music.

67. Sergiù Celibidache, 1912–1996, was a Romanian-born conductor highly regarded for his interpretations of the Romantic period. He taught many students and pinned his musical philosophy on the principles of phenomenology established by Edmund Husserl (1859–1938).

68. Nancy Carlson, *Harriet and the Garden* Carlson (Minneapolis, MN: Carolrhoda Books, 1997).

69. Robert Jourdain, *Music, the Brain and Ecstasy* (see note 34), 331.

70. Sheryl Kornman, "Grades Hit Higher Note: Music, Better Learning Linked in TUSD Study," *Tucson Citizen*, November 30, 2003: A1. Copyright 2003 *Tucson Citizen*. Reprinted with permission.

71. "OMA—Opening Minds Through the Arts—Evaluation Results," published online 2003 at http://www.omaproject.org/evaluation.htm (accessed May 6, 2004).

72. Parkview Elementary School, High Point, NC. North Carolina Department of Public Instruction: End-of-Grade tests;

statistics sent to school by Guilford County Schools and made available by Alan Parker, principal.

73. F. H. Rauscher, M. T. LeMieux, M. L. Kabele, and S. C. Hinton, "Lasting Improvement of At-Risk Children's Cognitive Abilities Following Music Instruction," manuscript submitted for publication. First presented as "Piano, Rhythm, and Singing Instruction Improve Different Aspects of Spatial-Temporal Reasoning in Head Start Children," poster presented at the annual meeting of the Cognitive Neuroscience Society, April 2003, New York.

Isabelle Peretz and her colleagues at the University of Montreal are also examining singing and the relationship of music and words in terms of neuroscience. See S. Hébert, A. Racette, L. Gagnon, and I. Peretz, "Revisiting the Dissociation between Singing and Speaking in Expressive Aphasia," *Brain* 126 (2003): 1838–1850. Also, I. Peretz, M. Radeau, and M. Arguin, "Two-Way Interactions Between Music and Language: Evidence from Priming Recognition of Tune and Lyrics in Familiar Songs," *Memory & Cognition*, in press; and I. Peretz, L. Gagnon, S. Hébert, and J. Macoir, "Singing in the Brain: Insights from Cognitive Neuropsychology," *Music Perception*, paper submitted for publication.

74. A useful resource is *Acts of Achievement: The Role of Performing Arts Centers in Education*, Barbara Rich, Jane L. Polin, and Stephen J. Marcus, eds. (Washington, DC: Dana Press, 2003), also available online in PDF format at www.dana.org.

75. Some suggestions for woodwinds: *Ensemble Repertoire for Woodwind Quintet*, Himie Voxman and Richard Hervig, eds., Rubank; *The Ross Taylor Woodwind Quintets*, Ross Taylor, arr., Southern Music Company; *Twenty Two Woodwind Quintets*, compiled and revised by Albert J. Andraud, Southern Music Company. There are many more publications available for string quartets, but broad compilations are rarer or out of print

(for example, the Flonzaly Quartet collection and the *Simply Strings* six-volume collection).

76. Stephen Schwartz, "The Colors of the Wind," *Pocahontas,* directed by Eric Goldberg and Mike Gabriel (Burbank, CA: Buena Vista Motion Picture Group, 1995).

SUGGESTIONS FOR FURTHER READING

Bowles, Shirley. "Tune Up the Mind: The Effect of Orchestrating Music as a Reading Intervention," Ed.D. diss. Indiana University of Pennsylvania, 2003.

Calvin, William H., *The Cerebral Symphony: Seashore Reflections on the Structure of Consciousness.* New York: Bantam New Age, 1990.

——, *How Brains Think: Evolving Intelligence, Then and Now.* New York: Basic Books, 1996.

Carter, Rita. *Mapping the Mind.* Berkeley, CA: University of California Press, 1999.

Cook, Deryck. *The Language of Music.* Oxford: Oxford Paperbacks, 1959.

DeArmond, Stephen J., Madeline M. Fusco, and Maynard M. Dewey. *Structure of the Human Brain: A Photographic Atlas.* Oxford: Oxford University Press, 1989.

Deutsch, Diana, ed. *The Psychology of Music.* New York: The Academic Press, second edition, 1999.

Diamond, M. C., A. Scheibel, and E. M. Elson, *The Human Brain Coloring Book.* New York: Harper Collins, 1985.

Flesch, Rudolph. *Why Johnny Can't Read.* Reprint, Perennial, 1986.

Freeman, Walter J. *Societies of Brains: A Study in the Neuroscience of Love and Hate.* Hillsdale, NJ: Lawrence Erlbaum, 1995.

Fuster, Joaquin M. *Cortex and Mind: Unifying Cognition.* Oxford: Oxford University Press, 2003.

Gardner, Howard. *Frames of Mind.* New York: Basic Books, 1985.

Healy, Jane M. *Endangered Minds: Why Children Don't Think and What We Can Do About It*. New York: Simon and Schuster, 1990.

Howard, Pierce J. *The Owner's Manual for the Brain*. Atlanta: Bard Press, 2000.

Johnson, Richard T., ed. *Neurology and Neuroscience: An Internet Resource Guide*. eMedguides.com, 2002.

Jourdain, Robert. *Music, the Brain and Ecstasy: How Music Captures Our Imagination*. New York: William Morrow, 1997.

LeDoux, Joseph. *Synaptic Self: How Our Brains Become Who We Are*. New York: Viking, 2002.

Leeuwen, Theo van. *Speech, Music, Sound*. New York: St. Martin's Press, 1999.

McGuinness, Diane. *Why Our Children Can't Read*. New York: Free Press, 1997.

Peretz, Isabelle, and Robert Zatorre, eds. *The Cognitive Neuroscience of Music*. Oxford: Oxford University Press, 2003.

Pinker, Steven. *How the Mind Works*. New York: W. W. Norton, 1997.

——, *The Language Instinct: How the Mind Creates Language*. New York: Harper Perennial, 1994.

Plato, *The Republic*. Trans. F. M. Cornford. Oxford: Oxford University Press, 1945.

Prelutsky, Jack, and Marilyn Hafner (illustrator). "The Turkey Shot Out of the Oven." In *It's Thanksgiving*. New York: Harper Collins, 1982.

Ratey, John J. *A User's Guide to the Brain*. New York: Pantheon Books, 2001.

Restak, Richard M. *The Brain*. New York: Bantam Books, 1984.

Shaw, Gordon L. *Keeping Mozart in Mind*. San Diego: Academic Press, 2000.

Springer, Sally P., and Georg Deutsch. *Left Brain, Right Brain: Perspectives from Cognitive Neuroscience*. New York: W. H. Freeman, 1999.

Tovey, Sir Donald. *The Forms of Music*. New York: Meridian Books, 1956.

Wolf, Maryanne, ed. *Dyslexia, Fluency and the Brain*. Timonium, MD: York Press, 2001.

INDEX

A

B

C

D

E

F

N

O

P

R

S

T

U

V

W

Y

OTHER DANA PRESS BOOKS AND PERIODICALS

www.dana.org/books/press

BOOKS FOR GENERAL READERS

THE CREATING BRAIN: The Neuroscience of Genius
Nancy C. Andreasen, Ph.D., M.D.
Andreasen, a noted psychiatrist and bestselling author, explores how the brain achieves creative breakthroughs—in art, literature, and science—including questions such as how creative people are different and the difference between genius and intelligence. She also describes how to nurture and develop our creative capacity. 33 illustrations/photos. 225 pp.

1-932594-07-8 • $23.95

THE ETHICAL BRAIN
Michael S. Gazzaniga, Ph.D.
Explores how the lessons of neuroscience help resolve today's ethical dilemmas, ranging from when life begins to "off-label" use of drugs such as Ritalin by students preparing for exams, and other topics, from free will and personal responsibility to public policy and religious belief. The author, a pioneer in cognitive neuroscience, is a member of the President's Council on Bioethics. 225 pp.

1-932594-01-9 • $25.00

FATAL SEQUENCE: The Killer Within
Kevin J. Tracey, M.D.
An easily understood account of the spiral of sepsis, a sometimes fatal crisis that most often affects patients fighting off nonfatal illnesses or injury. Tracey puts the scientific and medical story of sepsis in the context of his battle to save a burned baby, a sensitive telling of cutting-edge science. 225 pp.

Cloth, 1-932594-06-X • $23.95
Paper, 1-932594-09-4 • $12.95

A GOOD START IN LIFE: Understanding Your Child's Brain and Behavior from Birth to Age 6
Norbert Herschkowitz, M.D., and Elinore Chapman Herschkowitz
Updated with the latest information and new material, the authors show how young children learn to live together in family and society and how brain development shapes a child's personality and behavior, discussing appropriate rule-setting, the child's moral sense, temperament, language, playing, aggression, impulse control, and empathy.

Cloth 283 pp. 0-309-07639-0 • $22.95
Paper (Updated version with 13 illustrations) 312 pp.
0-9723830-5-0 • $13.95

BACK FROM THE BRINK: How Crises Spur Doctors to New Discoveries about the Brain
Edward J. Sylvester
In two academic medical centers, Columbia's New York Presbyterian and Johns Hopkins Medical Institutions, a new breed of doctor, the neurointensivist, save patients with life-threatening brain injuries. 16 illustrations/photos. 296 pp.

0-9723830-4-2 • $25.00

THE BARD ON THE BRAIN: Understanding the Mind Through the Art of Shakespeare and the Science of Brain Imaging
Paul Matthews, M.D., and Jeffrey McQuain, Ph.D. Foreword by Diane Ackerman
Explores the beauty and mystery of the human mind and the workings of the brain, following the path the Bard pointed out in 35 of the most famous speeches from his plays. 100 illustrations. 248 pp.

0-9723830-2-6 • $35.00

STRIKING BACK AT STROKE: A Doctor-Patient Journal
Cleo Hutton and Louis R. Caplan, M.D.
A personal account with medical guidance for anyone enduring the changes that a stroke can bring to a life, a family, and a sense of self. 15 illustrations. 240 pp.

0-9723830-1-8 • $27.00

THE DANA GUIDE TO BRAIN HEALTH
Floyd E. Bloom, M.D., M. Flint Beal, M.D., and David J. Kupfer, M.D., Editors. Foreword by William Safire
A home reference on the brain edited by three leading experts collaborating with 104 distinguished scientists and medical professionals. In easy-to-understand language with cross-references and advice on 72 conditions such as autism, Alzheimer's disease, multiple sclerosis, depression, and Parkinson's disease. 16 full-color pages and more than 200 black-and-white illustrations. 768 pp.

0-7432-0397-6 • $45.00

UNDERSTANDING DEPRESSION: **What We Know and What You Can Do About It**
J. Raymond DePaulo Jr., M.D., and Leslie Alan Horvitz.
Foreword by Kay Redfield Jamison, Ph.D.
What depression is, who gets it and why, what happens in the brain, troubles that come with the illness, and the treatments that work.

Cloth 304 pp. 0-471-39552-8 • $24.95
Paper 296 pp. 0-471-43030-7 • $14.95

KEEP YOUR BRAIN YOUNG: **The Complete Guide to Physical and Emotional Health and Longevity**
Guy McKhann, M.D., and Marilyn Albert, Ph.D.
Every aspect of aging and the brain: changes in memory, nutrition, mood, sleep, and sex, as well as the later problems in alcohol use, vision, hearing, movement, and balance.

Cloth 304 pp. 0-471-40792-5 • $24.95
Paper 304 pp. 0-471-43028-5 • $15.95

THE END OF STRESS AS WE KNOW IT
Bruce McEwen, Ph.D., with Elizabeth Norton Lasley. Foreword by Robert Sapolsky
How brain and body work under stress and how it is possible to avoid its debilitating effects.

Cloth 239 pp. 0-309-07640-4 • $27.95
Paper 262 pp. 0-309-09121-7 • $19.95

IN SEARCH OF THE LOST CORD: Solving the Mystery of Spinal Cord Regeneration
Luba Vikhanski
The story of the scientists and science involved in the international scientific race to find ways to repair the damaged spinal cord and restore movement. 21 photos; 12 illustrations. 269 pp.

0-309-07437-1 • $27.95

THE SECRET LIFE OF THE BRAIN
Richard Restak, M.D. Foreword by David Grubin
Companion book to the PBS series of the same name, exploring recent discoveries about the brain from infancy through old age. 201 pp.

0-309-07435-5 • $35.00

THE LONGEVITY STRATEGY: How to Live to 100 Using the Brain-Body Connection
David Mahoney and Richard Restak, M.D. Foreword by William Safire
Advice on the brain and aging well.

Cloth 250 pp. 0-471-24867-3 • $22.95
Paper 272 pp. 0-471-32794-8 • $14.95

STATES OF MIND: New Discoveries about How Our Brains Make Us Who We Are
Roberta Conlan, Editor
Adapted from the Dana/Smithsonian Associates lecture series by eight of the country's top brain scientists, including the 2000 Nobel laureate in medicine, Eric Kandel.

Cloth 214 pp. 0-471-29963-4 • $24.95
Paper 224 pp. 0-471-39973-6 • $18.95

THE DANA FOUNDATION
SERIES ON NEUROETHICS

HARD SCIENCE, HARD CHOICES: Facts, Ethics, and Policies Guiding Brain Science Today
Sandra Ackerman
This book, the newest in the Dana Foundation Series on Neuroethics, is based on an invitational meeting co-sponsored by the Library of Congress, the National Institutes of Health, the Columbia University Center for Bioethics, and the Dana Foundation. Top scholars and scientists discuss new and complex medical and social ethics brought about by advances in neuroscience. 200 pp.
1-932594-02-7 • $12.95

NEUROSCIENCE AND THE LAW: Brain, Mind, and the Scales of Justice
Brent Garland, Editor. Foreword by Mark S. Frankel. With commissioned papers by Michael S. Gazzaniga, Ph.D., and Megan S. Steven; Laurence R. Tancredi, M.D., J.D.; Henry T. Greely, J.D.; and Stephen J. Morse, J.D., Ph.D.
How discoveries in neuroscience influence criminal and civil justice, based on an invitational meeting of 26 top neuroscientists, legal scholars, attorneys, and state and federal judges convened by the Dana Foundation and the American Association for the Advancement of Science. 226 pp.
1-932594-04-3 • $8.95

BEYOND THERAPY: Biotechnology and the Pursuit of Happiness. A Report of the President's Council on Bioethics
Special Foreword by Leon R. Kass, M.D., Chairman. Introduction by William Safire
Can biotechnology satisfy human desires for better children, superior performance, ageless bodies, and happy souls? This report says these possibilities present us with profound ethical challenges and choices. Includes dissenting commentary by scientist members of the Council. 376 pp.
1-932594-05-1 • $10.95

NEUROETHICS: Mapping the Field. Conference Proceedings.
Steven J. Marcus, Editor
Proceedings of the landmark 2002 conference organized by Stanford University and the University of California, San Francisco, at which more than 150 neuroscientists, bioethicists, psychiatrists and psychologists, philosophers, and professors of law and public policy debated the implications of neuroscience research findings for individual and societal decision-making. 50 illustrations. 367 pp.

0-9723830-0-X • $10.95

FREE EDUCATIONAL BOOKS

(Information about ordering and downloadable PDFs are available at www.dana.org.)

PARTNERING ARTS EDUCATION: A Working Model from ArtsConnection
This publication illustrates the importance of classroom teachers and artists learning to form partnerships as they build successful residencies in schools. Partnering Arts Education provides insight and concrete steps in the ArtsConnection model. 55 pp.

ACTS OF ACHIEVEMENT: The Role of Performing Arts Centers in Education.
Profiles of 60-plus programs, plus eight extended case studies, from urban and rural communities across the United States, illustrating different approaches to performing arts education programs in school settings. Black-and-white photos throughout. 164 pp.

PLANNING AN ARTS-CENTERED SCHOOL: A Handbook
A practical guide for those interested in creating, maintaining, or upgrading arts-centered schools. Includes curriculum and development, governance, funding, assessment, and community participation. Black-and-white photos throughout. 164 pp.

THE DANA SOURCEBOOK OF BRAIN SCIENCE: Resources for Secondary and Post-Secondary Teachers and Students
A basic introduction to brain science, its history, current understanding of the brain, new developments, and future directions. 16 color photos; 29 black-and-white photos; 26 black-and-white illustrations. 160 pp.

THE DANA SOURCEBOOK OF IMMUNOLOGY: Resources for Secondary and Post-Secondary Teachers and Students
An introduction to how the immune system protects us, what happens when it breaks down, the diseases that threaten it, and the unique relationship between the immune system and the brain. 5 color photos; 36 black-and-white photos; 11 black-and-white illustrations. 116 pp.

ISSN: 1558-6758

PERIODICALS

Dana Press also offers several periodicals dealing with arts education, immunology, and brain science. These periodicals are available free to subscribers by mail. Please see www.dana.org.

PETER PERRET is the Conductor Emeritus of the Winston-Salem Symphony. He consults with school boards across the nation and supervises music education research in conjunction with the Section of Neuropsychology at Wake Forest University Health Sciences. He is a frequent guest lecturer at science and education conferences dealing with music and learning.

JANET FOX is a freelance arts and education writer based in Winston-Salem, N.C. She has twice won the North Carolina Association of Educators' School Bell Award for education reporting. She is the author, with Christie Latona, of *The Playful Power of Metaphor: Harness the Winds of Creativity, Innovation and Possibility* (2005).